Łukasz Żwirełło

Realization Limits of Impulse-Radio UWB Indoor Localization Systems

Karlsruher Forschungsberichte
aus dem Institut für Hochfrequenztechnik und Elektronik

Herausgeber: Prof. Dr.-Ing. Thomas Zwick

Band 71

Realization Limits of Impulse-Radio UWB Indoor Localization Systems

by
Łukasz Żwirełło

Dissertation, Karlsruher Institut für Technologie (KIT)
Fakultät für Elektrotechnik und Informationstechnik, 2013

Impressum

Scientific
Publishing

Karlsruher Institut für Technologie (KIT)
KIT Scientific Publishing
Straße am Forum 2
D-76131 Karlsruhe

KIT Scientific Publishing is a registered trademark of Karlsruhe
Institute of Technology. Reprint using the book cover is not allowed.

www.ksp.kit.edu

Print on Demand 2013

ISSN 1868-4696
ISBN 978-3-7315-0114-5

Vorwort des Herausgebers

Anfänglich motiviert durch die permanent steigende Nachfrage nach höheren Datenraten im Bereich der mobilen Kommunikation wird seit einigen Jahren intensiv an einer neuen Technologie geforscht: der Ultrabreitband-Technik (UWB). Die Sendesignale werden hierbei über eine sehr große Bandbreite gespreizt, was zu einer sehr geringen spektralen Leistungsdichte und damit einem sehr geringen Störpotential für andere Funkdienste führt. Die vor einigen Jahren in vielen Zonen der Erde dazu freigegebenen Frequenzbereiche ebnen den Weg für zukünftige kommerzielle UWB-Systeme und haben die Forschungstätigkeit weiter angefacht. Die ursprüngliche Idee, die neue Technologie zur übertragung sehr großer Datenraten über kurze Strecken zu nutzen ist mittlerweile allerdings nicht mehr im Fokus der UWB-Forschung. Stattdessen werden Sensorikanwendungen oder z.B. die Lokalisierung innerhalb von Gebäuden anvisiert.

Satellitengestützte Lokalisierungssysteme bieten nur Positionsgenauigkeiten im Bereich eines Meters und verlieren noch weiter an Genauigkeit, wenn die Sichtverbindung zu den Satelliten blockiert wird, wie es innerhalb von Gebäuden generell der Fall ist. Hier werden UWB-Lokalisierungssysteme zur attraktiven Alternative, da die extrem große Bandbreite hoch genaue Entfernungsmessungen ermöglicht. Insbesondere Pulsbasierte Systeme mit ihrem Potential für eine einfache und kostengünstige Implementierung erscheinen hierbei als besonders vorteilhaft, da bei der Lokalisierung im Gegensatz zu typischen Radaranwendungen keine Auflösung derMehrwegeausbreitung erreicht, sondern nur die Laufzeit der direkten Verbindung gemessen werden muss. Bei zusätzlicher Codierung der Pulse lassen sich sogar gleichzeitig Daten übertragen.

In seiner Dissertation untersuchte Herr Żwirełło die Grenzen der Realisierbarkeit von impulsbasierten UWB Lokalisierungssystemen. Dazu hat er mehrere Sende- und Empfangsmodule von der Antenne bis zum Basisband (inkl. FPGA) erfolgreich realisiert und daraus ein komplettes Lokalisierungssystem gebildet. Damit konnte erstmalig ein Vergleich der drei wichtigsten UWB Transceiver-Architekturen, inkohärente Auto-Korrelation sowie kohärente Korrelation mit Referenzpuls und Transmitted Reference hinsichtlich Komplexität und Performanz erfolgen. Herr Żwireło führte erfolgreich eine theoretische Bestimmung der Lokalisierungsgenauigkeit und Verifikation anhand von Messungen in unterschiedlichen Szenarien durch.

Die vorliegende Arbeit von Herrn Żwireło stellt eine wesentliche Grundlage für künftige Forschung auf dem Gebiet der UWB basierten Lokalisierung dar. Die Ergebnisse werden weltweit Aufmerksamkeit auf sich ziehen und weitere Forschungsaktivitäten auf diesem Gebietmit sich bringen. Ich wünsche Herrn Żwireło für seine Zukunft weiterhin viel Erfolg und bin mir sicher, dass er noch viele wesentliche Beiträge zu Wissenschaft und Technik leisten wird.

Prof. Dr.-Ing. Thomas Zwick
– Institutsleiter –

Forschungsberichte aus dem
Institut für Höchstfrequenztechnik und Elektronik (IHE)
der Universität Karlsruhe (TH) (ISSN 0942-2935)

Herausgeber: Prof. Dr.-Ing. Dr. h.c. Dr.-Ing. E.h. mult. Werner Wiesbeck

Forschungsberichte aus dem
Institut für Höchstfrequenztechnik und Elektronik (IHE)
der Universität Karlsruhe (TH) (ISSN 0942-2935)

Forschungsberichte aus dem
Institut für Höchstfrequenztechnik und Elektronik (IHE)
der Universität Karlsruhe (TH) (ISSN 0942-2935)

Forschungsberichte aus dem
Institut für Höchstfrequenztechnik und Elektronik (IHE)
der Universität Karlsruhe (TH) (ISSN 0942-2935)

Karlsruher Forschungsberichte aus dem
Institut für Hochfrequenztechnik und Elektronik
(ISSN 1868-4696)

Herausgeber: Prof. Dr.-Ing. Thomas Zwick

Die Bände sind unter www.ksp.kit.edu als PDF frei verfügbar
oder als Druckausgabe bestellbar.

Karlsruher Forschungsberichte aus dem
Institut für Hochfrequenztechnik und Elektronik
(ISSN 1868-4696)

Band 63   Małgorzata Janson
          **Hybride Funkkanalmodellierung für ultrabreitbandige
          MIMO-Systeme** (2011)
          ISBN 978-3-86644-639-7

Band 64   Mario Pauli
          **Dekontaminierung verseuchter Böden durch
          Mikrowellenheizung** (2011)
          ISBN 978-3-86644-696-0

Band 65   Thorsten Kayser
          **Feldtheoretische Modellierung der Materialprozessierung
          mit Mikrowellen im Durchlaufbetrieb** (2011)
          ISBN 978-3-86644-719-6

Band 66   Christian Andreas Sturm
          **Gemeinsame Realisierung von Radar-Sensorik und
          Funkkommunikation mit OFDM-Signalen** (2012)
          ISBN 978-3-86644-879-7

Band 67   Huaming Wu
          **Motion Compensation for Near-Range Synthetic Aperture
          Radar Applications** (2012)
          ISBN 978-3-86644-906-0

Band 68   Friederike Brendel
          **Millimeter-Wave Radio-over-Fiber Links based on
          Mode-Locked Laser Diodes** (2013)
          ISBN 978-3-86644-986-2

Band 69   Lars Reichardt
          **Methodik für den Entwurf von kapazitätsoptimierten
          Mehrantennensystemen am Fahrzeug** (2013)
          ISBN 978-3-7315-0047-6

Band 70   Stefan Beer
          **Methoden und Techniken zur Integration von 122 GHz
          Antennen in miniaturisierte Radarsensoren** (2013)
          ISBN 978-3-7315-0051-3

Band 71   Łukasz Żwirełło
          **Realization Limits of Impulse-Radio UWB Indoor
          Localization Systems** (2013)
          ISBN 978-3-7315-0114-5

Realization Limits of Impulse-Radio UWB Indoor Localization Systems

Zur Erlangung des akademischen Grades eines

DOKTOR-INGENIEURS

von der Fakultät für
Elektrotechnik und Informationstechnik
des Karlsruher Instituts für Technologie (KIT)

genehmigte

DISSERTATION

von

Dipl.-Ing. Łukasz Żwirełło
aus Gdańsk, Polen

Tag der mündlichen Prüfung: 27. Juni 2013
Hauptreferent: Prof. Dr.-Ing. Thomas Zwick
Korreferent: Prof. Dr.-Ing. Martin Vossiek

Acknowledgements

This dissertation resulted from my work as research and teaching assistant at the Institut für Hochfrequenztechnik und Elektronik (IHE) at the Karlsruhe Institute of Technology (KIT).

First and foremost, I would like to express my deep sense of gratitude to my supervisor, Professor Thomas Zwick. He offered me his time for long and detailed discussions in the initial phase of my work and at the same time left plenty of scientific freedom to follow my own ideas. Thank you for your support - not only in technical matters. I gratefully acknowledge Professor Martin Vossiek of the Friedrich-Alexander-Universität Erlangen-Nürnberg for agreeing to be the *Korreferent* of my thesis and the member of the examination committee, as well as for the valuable comments.

Furthermore I would like to thank the Professor Hermann Schumacher of the Institut für Elektronische Bauelemente und Schaltungen at the Universität Ulm for providing the Ultra-Wideband ICs as well as the essential hints. Moreover, my gratitude goes to Professor Jürgen Fleischer of the Institut für Produktionstechnik as well as to Professor Gert F. Trommer of the Institut für Theoretische Elektrotechnik und Systemoptimierung for particularly fruitful cooperation in the field of industrial localization and pedestrian navigation respectively.

To all my colleagues at IHE, many thanks for your friendly appearance, full of helpfulness and for unforgettable time I spend in Karlsruhe. Special thanks also go to my office colleagues, who I am lucky to call my friends now: Dr.-Ing. Jens Timmermann, Dr.-Ing. Grzegorz Adamiuk, Christoph Heine and Tom Schipper (especially for your endless patience in correcting almost countless documents I have written in German). I am particularly indebted to Dr.-Ing. Friederike Brendel and Dr.-Ing. Christian Sturm for proofreading the manuscript with impressive accuracy and patience, as well as to Leen Sit - colleague and the very first student I had the opportunity to supervise - for grammar improvements.

Further thanks go to my friends, who supported me in the key moments and provided healthy portion of distraction in all other ones.

I thank my parents, Maria and Bogdan, for "being there" for me at any time
and for their love, which made me strong enough to deal with ups and downs
I have ensountered in the past three decades - *Dziękuję Wam kochani rodzice!*
They gave me also the best brother I could imagine. There is no stronger bond
than the "brotherhood" - it gave me the energy for doing extraordinary things.

Last but not least, I thank my beloved wife Monika for her support and having
my back every time I needed it. She is my biggest inspiration!

Karlsruhe, in November 2013
Łukasz Żwirełło

Abstract

In this thesis, the realization limits of an impulse-based Ultra-Wideband (UWB) localization system for indoor applications have been thoroughly investigated and verified by measurements. The analysis spans from the position calculation algorithms, through hardware realization and modeling, up to the localization experiments conducted in realistic scenarios. The main focus was put on identification and characterization of limiting factors as well as developing methods to overcome them.

Thanks to the wide application of Global Positioning System (GPS) receivers in everyday use electronic devices, the nearly continuous knowledge of own location has become an almost obvious thing for their users. However, in spite of the significant interest, there is still no counterpart for GPS indoors. The UWB technology, using short electromagnetic impulses, deals very well with the multipath rich indoor environments and does not undergo fading, being a perfect candidate for use in an accurate localization system. The general idea of UWB localization is to calculate the position of a mobile, impulse-transmitting tag based on the signal time of flight measurements to multiple reference points.

The system analysis begins with the determination of algorithms offering the best trade-off between time efficiency and precision. Among investigated iterative approaches, a non-iterative algorithm for solving the localization equations is introduced in this thesis, which significantly reduces the computation time. Furthermore, the suitable error monitoring methods for use in indoor positioning have been identified and implemented.

In terms of the achievable operation range the selection of optimal receiver architecture is a crucial issue during UWB system design. In this work a new comparison metric, called signal-to-threshold ratio (STR), is introduced. It allows comparison of impulse-based systems directly in time domain. For the first time the three most common UWB receiver topologies were compared in measurements, while using a self-designed common hardware platform.

Furthermore, the coherent UWB receiver has been investigated, as it has the best performance at large distances and low signal power. In this work the synchronization and impulse tracking schemes were developed, implemented on an FPGA and tested with UWB hardware for the first time. The performed comprehensive parameter study enabled quantification of the sensitivity of this transmission technique with respect to timing issues and determination of the most reliable synchronization method. Moreover, it has been demonstrated that picosecond-precise and fast-adjustable clock signals for use in UWB correlation receivers can be realized with a phase-locked-loop and an adjacent programmable delay element.

Another step towards better understanding of the localization accuracy and precision limits was to investigate transmitter and receiver hardware components, as well as the time measurement device. The performance degrading effects associated with the angle-dependent antenna behavior and threshold-based signal detection have been identified and the correction methods have been proposed and verified. In addition, based on the error models acquired, the Cramer-Rao lower bound (CRLB) has been derived for this 3D time-difference of arrival localization system. The CRLB can be used as the benchmark for the localization system precision lower limit. The performed localization tests have proven the correctness of the accuracy and the precision estimation methods, both depending on the geometrical configuration of the system. The average horizontal (2D) and 3D accuracy obtained during experiments conducted in an industrial-like environment are 3.4 cm and 11.4 cm, respectively. The comparison with the literature published up to date proves the excellent quality of these results.

The methods of systematic localization system analysis, reaching from algorithms, over hardware, up to measurements, have been developed in this work, and enable undertaking of aware and targeted improvements. The given guidelines can be used for design and optimization of future application-tailored positioning systems and are not limited to UWB technology only.

Zusammenfassung

Die Ultra-Breitband (engl. Ultra-Wideband UWB) Technologie ist eine relativ junge und noch wenig erforschte Funkübertragungstechnik. Bei ihr kommen kurze elektromagnetische Impulse mit geringer mittlerer Leistung zum Einsatz. Dank der großen Bandbreite der verwendeten Signale unterliegen sie nicht dem Mehrwegeschwund. Aus diesem Grund bietet sich der Einsatz dieser Technologie primär für störungsresistente Kommunikations- und präzise Ortungssystemen an.

In der vorliegender Arbeit werden die Realisierbarkeitsgrenzen eines impulsbasierten UWB Ortungssystems für Indoor-Anwendungen untersucht und messtechnisch verifiziert. Die Analyse reicht von der Wahl der Positionsberechnungsalgorithmen, über die Hardwareentwicklung, bis hin zu Lokalisierungsexperimenten, die in realistischen Szenarien durchgeführt werden. Die Identifizierung und Charakterisierung performanzlimitierender Faktoren, sowie das Entwickeln von Ansätzen um diese umzugehen, liegen dabei im Fokus.

Die Betrachtung des Systems beginnt mit der Analyse der Algorithmen, die eine gute Zeiteffizienz aufweisen und präzise Lösungen liefern. Es werden dabei sowohl iterative als auch nichtiterative Lösungsansätze untersucht. Des Weiteren werden geeignete Fehler-Monitoring Methoden für den Einsatz in der UWB Lokalisierung identifiziert und implementiert.

Im Hinblick auf die maximal erreichbare Übertragungsdistanz spielt die Wahl der Empfängerarchitektur eine entscheidende Rolle. In der vorliegenden Arbeit wird eine neue Vergleichsmetrik definiert, die es erlaubt, die Performanz der UWB-Empfänger direkt im Zeitbereich zu ermitteln. Die theoretischen Untersuchungen werden durch Simulationen und Messungen verifiziert und bestätigt. Die dazu benötigte Hardware wird ebenfalls in dieser Arbeit entworfen.

Ausgehend von der obigen Studie, wird ein Korrelationsempfänger als effizienteste Lösung identifiziert und genauer untersucht. Hierfür werden Pulssynchronisations- und Pulstrackingalgorithmen entworfen, auf einem FPGA implementiert und zusammmen mit der UWB-Hardware getestet. Die durchgeführte Parameterstudie erlaubt eine Aussage über die Empfindlichkeit eines solchen Korrelationsempfängers gegenüber der Synchronisationsgenauigkeit. Des Weiteren wird gezeigt, dass pikosekundengenaue und schnell anpassbare Zeitsignale mit Hilfe einer Phasenregelschleife und programmierbaren Verzögerungselementen realisiert werden können.

Ein weiterer Schritt, der dazu führte die Lokalisierungsgenauigkeit und -präzision des Systems besser einschätzen zu können, ist eine Fehleranalyse der einzelnen Komponenten. Die dabei gewonnenen quantitativen Kenntnisse über die Nichtidealitäten des Systems erlauben die Herleitung der unteren Cramer-Rao Schranke (CRLB). Die CRLB beschreibt das untere Limit der Lokalisierungspräzision und wird hier erstmalig für ein 3D zeitdifferenzbasiertes Ortungssystem hergeleitet. Die durchgeführten Ortungsexperimente bestätigen die vorgeschlagenen Methoden zur Abschätzung der Genauigkeit und Präzision, welche beide von der geometrischen Systemanordnung abhängen. Die durchschnittliche horizontale (2D) und 3D Genauigkeit bei Lokalisierungsversuchen in einer industrienahem Umgebung beträgt 3.4 cm bzw. 11.4 cm. Der Vergleich mit Werten aus bis Dato veröffentlichter Literatur bestätigt die hervorragende Qualität dieser Resultate.

In dieser Arbeit wurde eine systematische Analyse eines impulsbasierten Lokalisierungssystem durchgeführt, ausgehen von Algorithmen, über Hardware, bis hin zu Messungen. Die dabei entwickelten Methoden erlauben gezielte Systemeingriffe mit genau prädizierbarem Resultaten. Die entwickelten Richtlinien können künftig beim Design und der Optimierung von anwendungsangepassten Lokalisierungssystemen eingesetzt werden und beschränken sich dabei nicht ausschließlich für die UWB Technologie.

Table of contents

Acronyms and symbols

Lower case letters

c_0	speed of light in vacuum ($\approx 2.997925 \cdot 10^8$ m/s)
d_{ab}	distance between point a and b
e	template signal (time domain)
f	frequency
f_L, f_H	lower and upper cut-off frequency
h	antenna impulse response
n	noise signal (time domain)
p	pulse shape in time domain
r	received signal (time domain)
$\vec{r}$	position vector
s	signal (time domain)
t	time
$t_{a,b}$	signal time of flight between points a and b
x,y,z	Cartesian coordinates
w	time measurement noise

Capital letters

A	signal amplitude
C	covariance matrix
D	decorrelated covariance matrix (diagonal)
D	sub-function of the FIM, related to a certain coordinate
$\mathcal{F}$	Fourier transformation
$\mathcal{F}^{-1}$	inverse Fourier transformation
G	antenna gain
H	measurement / mapping matrix
H	antenna transfer function
I	Fisher Information Matrix

$\mathbf{I}$	identity matrix
$\mathbf{L}$	lower (left) triangular matrix
$\mathcal{N}$	normal distribution
P	power of a signal
R	range / distance
S	pulse spectrum
T	period (time)
$\mathbf{U}$	upper (right) triangular matrix
Z_S	system impedance

Mathematical operators and indexes

$\square^{\mathrm{T}}$	transposed matrix
$\|\dots\|$	Euclidean norm of the vector
$cov\{\dots\}$	covariance matrix of the vector
$E\{\dots\}$	expected value
$p(\dots)$	probability function
$\square_S$	transmitter position
$\square_{E,i}$	position of the i-th receiver
$\square_{ij}$	measurement difference between i-th and j-th receiver
$\hat{\square}$	estimated real value
$\tilde{\square}$	measured real value

Greek symbols

$\Delta\square$	difference of two variables
η	efficiency
$\vec{\rho}$	vector of time or distance differences
ψ	azimuth angle
θ	elevation angle
$\sigma_\square$	standard deviation of a parameter

Acronyms

AC	Alternating Current
ACF	Autocorrelation Function

ACR   Auto-Correlation Receiver
ADC   Analog to Digital Conversion
AIR   Antenna Impulse Response
AOA   Angle of Arrival
AP   Access Point
APC   Antenna Phase Center
AWGN   Additive White Gaussian Noise
BA   Bancroft Algorithm
BPF   Bandpass Filter
BR   Base Receiver
CR   Correlation Receiver
CRLB   Cramer-Rao Lower Bound
DAC   Digital to Analog Conversion
DDS   Direct Digital Synthesis
DC   Direct Current
DOP   Dilution of Precision
ECC   Electronic Communications Committee
EIRP   Equivalent Isotropic Radiated Power
FA   Frequency Accuracy
FCC   Federal Communications Commission
FIM   Fisher Information Matrix
FMCW   Frequency Modulated Continuous Wave
FPGA   Field Programmable Gate Array
FWHM   Full Width at Half Maximum
GN   Gauss-Newton Algorithm
GPS   Global Positioning System
IC   Integrated Circuit
INS   Inertial Navigation System
IR   Impulse Radio
LAN   Local Area Network
LC   Loosely Coupled
LNA   Low Noise Amplifier
LM   Levenberg-Marquardt Algorithm
LOS   Line of Sight
LPF   Lowpass Filter
LUT   Look-Up Table
MU   Mobile User
NLOS   Non Line of Sight

NWA	Network Analyzer
P1dB	1 dB compression point
PAR	Peak-to-Average Power Ratio
PD	Programmable Delay
PDF	Probability Density Function
PDP	Power Delay Profile
PG	Pulse Generator
PL	Path Loss
PLL	Phase Locked Loop
PP	Pilot Pulse
PPM	Pulse Position Modulation
PRF	Pulse Repetition Frequency
PSD	Power Spectral Density
QACR	Quadrature Analog Correlation Receiver
RAIM	Receiver Autonomous Integrity Monitoring
RBW	Resolution Bandwidth
RCM	Range Comparison Method
RF	Radio Frequency
RSS	Received Signal Strength
RTLS	Real Time Localization System
SNR	Signal to Noise Ratio
SSR	Serial Shift Register
STR	Signal to Threshold Ratio
TDC	Time to Digital Converter
TDOA	Time Difference of Arrival
TJ	Time Jitter
TOA	Time of Arrival
TOF	Time of Flight
TP	Template Pulse
TR	Transmitted Reference Receiver
UWB	Ultra-Wideband
VCO	Voltage Controlled Oscillator
WLAN	Wireless Local Area Network
XT	Cross Talk

All other specific abbreviations will be introduced in the text.

1. Introduction

In the recent decade, during which global positioning system (GPS) receivers have been successively integrated in more and more devices dedicated for the consumer market (e.g. cellular phones, digital cameras or tablets), the nearly continuous knowledge of own location (position awareness) has become an almost obvious thing for many of us. The accuracy of GPS is, however, limited to tens of centimeters or even meters, which can be sufficient for applications used for navigating through a city or geo-tagging of photos taken with a camera. Could this be adopted for indoor scenarios as well? Firstly, in most cases the GPS signals are too weak to penetrate the structure of the building, and even if this was possible, they would be too distorted to be used for position calculation. Secondly, for most indoor applications the required accuracy is at least one order of magnitude higher than what the current GPS could offer.

The variety of indoor scenarios and applications that could profit from position awareness is countless. Safety, comfort, efficiency - all of those could be improved if only the position of diverse objects could be monitored in a remote way. In order to illustrate this better, let us consider the example of an enclosed environment where the requirements on the accuracy and reliability of a localization system are very high, such as factory buildings or assembly halls. The example of this kind of a scenario is depicted in Fig. 1.1.

In present industrial environments, the control of robots and production processes is performed over wires or fiber optics, whereas their position estimation is accomplished e.g. over inductive coupling with cables buried in the floor (proximity localization). The undiminished wish for more flexibility of the production lines, shorter manufacturing cycles and location aware, autonomous systems could be largely satisfied by the introduction of a technology allowing wireless position estimation. According to a survey conducted in [16] an accuracy of less than 10 cm is sufficient to enable a whole new range of applications. If parallel to the localization feature the data transmission would be additionally possible, the profit could be even larger.

With further consideration of the example from Fig. 1.1, in environments where people and machines have to cooperate within confined spaces a wireless traffic monitoring system could provide a new level of safety. For example instead of placing the robots behind safety fence to avoid injuries, they could be stopped when a person (equipped with a localizable tag) approaches them too closely. The mobile systems, such as forklifts, could be localized in real time and this information be incorporated in a collision avoidance system. With data transmission capabilities a wireless localization system could even allow for autonomous operation of driverless vehicles or even a large part of the entire production cycle. However, at this point the question arises: *which technology could fulfill all the above expectations?*

Fig. 1.1.: The use of the impulse-radio Ultra-Wideband technology, allowing for precise wireless localization with parallel data communication, can enable a whole new range of applications in challenging indoor environments.

Nowadays several wireless localization methods based on different technologies, targeted for various indoor environments are being investigated worldwide. They can be divided into acoustic, optical, and radio frequency (RF) methods. A few of those systems, along with their typical parameters, are listed in table 1.1. Based on this overview it becomes clear, that ultrasound could well achieve the accuracy of 10 cm or even less. On the other hand, the amount of required sensors is large [Pri05] and the possibility of data communication is very limited [ACH+01]. Additionally, a noisy environment can cause the localization performance to drop.

The optical systems, such as [Nor13], achieve superior localization performance at sub-millimeter level. Optical systems, however, require markers distributed in the scenario, in order to navigate and localize themselves. Hence, during the reconfiguration it is necessary to place new markers, and calibrate their reference positions. Additionally, the line of sight between the markers and the reader has to be maintained at all times. The data exchange with the infrastructure on an optical basis is only conditionally possible. Those draw-

Table 1.1.: Selected indoor localization systems.

System	Technology	Method	Accuracy	Remarks
ActiveBat [ACH$^+$01]	Ultrasonic	TDOA	< 10 cm	commercial
Cricket [Pri05]	Ultrasonic, Radio	proximity	10 cm	research
Ekahau [Eka13, SGAR07]	WLAN	RSS	1 m	commercial
WhereNet [Whe13]	WLAN	TDOA	2-3 m	commercial
Rover [BAK$^+$02]	WLAN + Bluetooth	RSS	2 m	research
PinPoint [YYR$^+$06]	RFID	TOA	1-3 m	research
Metro Group Future Store	RFID	TDOA + AOA	30 cm	—
Tadlys Topaz [Wei04]	Bluetooth + Infrared	CoO	2 m	commercial
Optotrak [Nor13]	optical / laser	—	< 0.5 mm	commercial

backs make optical systems unsuitable for the applications where adaptability to a new scenario geometry and possibility of data transmission are required.

Another localization method employs the RF signals. RF systems, mostly employing carrier-based, narrowband signals (e.g. IEEE 802.11 or RFID), are widely used in communication and proximity sensing. Their main advantage is the simplicity of deployment and reconfiguration. In the past decade several attempts have been undertaken to extend their functionality to enable localization, while using the same subsystems. The approach to reuse the well-established technology could save a lot time and money for hardware development. However, as presented in table 1.1, the accuracy of those systems is approximately one order of magnitude too low for them to find applications in e.g. industrial indoor navigation ([Eka13],[YYR$^+$06]). Additionally, the narrowband systems suffer from lack of immunity against multipath fading[1], leading to location-dependent break-downs in signal strength. An interesting approach has been presented in [RGV08], where an FMCW[2]-based ranging system, operating in the 5.8 GHz ISM[3] band, achieved a ranging accuracy

---

[1] fading - constructive and destructive superposition of multipath signal components, occurring periodically in space, leading to severe signal power fluctuations at the receiver
[2] FMCW - frequency-modulated continuous wave
[3] ISM - industrial, scientific and medical

better than 5 cm. The reported results give high hopes for achieving a 3D localization accuracy better than 10 cm. Although the fading is less of a problem in the FMCW architecture, the multipath signal components can however severely degrade the ranging performance. The authors of [RGV08] propose a method of mitigating this effect, however for this purpose a high signal bandwidth is required [WWG08]. Due to those difficulties and limitations, it is an extremely challenging task to establish reliable operation of accurate radio-based localization in indoor environments.

The potential solution could be the use of very short electromagnetic impulses. Such signals are resistant against multipath fading, due to their broadband character, and as such are ideal candidates for reliable indoor transmission. The data communication functionality can be achieved by simple modulation of the pulses (e.g. by changing their amplitudes or mutual distances). Their short time duration ensures at the same time a supreme time resolution which is a substantial advantage for indoor localization applications. The technology offering such signals is impulse-radio Ultra-Wideband (IR-UWB).

1.1. State of the art in IR-UWB systems and localization

The basic idea of Ultra-Wideband can be traced back to the experiments of Heinrich Hertz, performed in 1886 at the Technische Hochschule Karlsruhe, and those of Guglielmo Marconi from mid 1890s, using spark gap transmitters. The resulting very short impulses, that were spread over a wide frequency band were the basis of the first wireless communications. This system concept, however, was not adopted for radio communications in the long run due to the fact that each transmitter occupied too much of the available spectrum and the probability of interference between different users was extremely high. Instead, for the past century, radio communication has been performed over precisely assigned frequency bands.

In the early 1960s, the UWB technology was reborn, when attempts to adopt this technology for military use began. For the subsequent two decades, this technology was implemented in various radar and communication systems, profiting from the unique time resolution of short electromagnetic pulses, hence offering the possibility to precisely measure the distance. The energy of short pulses is spread over a large bandwidth in the spectral domain, resulting in low probability of detection.

One of the major limitations of UWB to enter the commercial market was the legislation. Since this technique occupies a wide frequency spectrum, it has to coexist with the conventional transmission systems, without causing any interference and performance degradation perceptible by their users. After several years of work, the Federal Communications Commission (FCC) in the US assigned a big portion of spectrum for license-free usage in 2002 [Fed02]; the Electronic Communications Committee (ECC) in Europe followed their footsteps in 2007 [Ele06]. These regulations strictly define the allowed transmit power and the frequency profile, making it possible for UWB devices to be operated as an overlay system.

In the first years, after the FCC issued the UWB regulation, a lot of attention was paid to the high-speed data transfer potential of this technology [RFSL04]. Many saw the UWB as the short range ($\approx 3\,\text{m}$) cable replacement, with the capability to transfer hundreds of Mbps [Str03]. Time has shown however, that Ultra-Wideband can be used more efficiently as localization system, with communication capabilities only as an additional feature.

Criteria for transceiver comparison
In the last years several UWB demonstrators for communication and localization purposes have been built [Sto08], [DSD$^+$07]. However, the choice of the transceiver architecture is based on the theoretical analysis of the advantages and disadvantages that can be found in the literature (e.g. [Ree05], ch. 6). Here the question of *"how to fairly compare the receiver topologies with each other?"* arises. An aspect often discussed in scientific publications considers the energy consumption issues of the receiver structures, mainly with regard to the placement of the A/D converter [VD08]. This includes the fully-, semi-digital and almost entirely analog receivers. Another, often chosen receiver comparison criterion is the simulated BER (or E_b/N_0) [DB05]. The most universal criterion, however, would be the evaluation of the signal-to-noise ratio in the baseband signal, before the A/D conversion. To the author's best knowledge, no research results describing such performance verification and backed up with measurement data, have been published up to date. All this type of considerations are either based on purely mathematical approaches [CB07] and do not take non-ideal effects into account, or the comparison is done under very specific assumptions (e.g. relative to rake receiver [QW05]), making it impossible to draw generalized conclusions.
Therefore, in this thesis a universal comparison criterion for evaluation of the SNR performance of IR-UWB receivers in time domain has been developed. Based on this, several receiver types have been realized using a common

hardware platform, in order to enable measurement and simulation based performance verification and confrontation with theory.

Optimal IR-UWB receiver

In many theoretical considerations the correlation receiver is said to be unconditionally the most efficient one in terms of SNR [WS97, Ree05, Tim10]. The main issue during coherent UWB transmission is the achieving of the time alignment between the received and template pulse, in order to maximize the correlation gain. In the past 15 years numerous algorithms have been developed to perform this task, where the synchronization speed was the main quality measure. However, while proposing an algorithm most of the authors tend to make very specific assumptions regarding either the radio channel (e.g. presence of dense multipath [HS02]), transmitted information (e.g. data aided synchronization [YG03]) or available hardware (e.g. parallel use of energy detection and correlation receiver [MZH08], or use of a range of template pulse shapes [WO06]). This suggests that a lot of them seem to have neglected the aspect of practical realizability. As a consequence only a hand full of practical realizations of the correlation receiver can be found in literature and even less demonstrate potential allowing to hope that they could be developed beyond the "proof of concept"-phase [TYM$^+$06].

Due to these reasons, there is nearly no information to be found regarding the synchronization stability in UWB correlation receivers and the effort required to achieve and maintain it under realistic conditions. Therefore, the realization and investigation of such a system is the second balance point of the work presented here.

Limitations in impulse-based localization

The lack of full understanding of the above described problems has its continuation when it comes to designing the high-performance IR-UWB localization systems. A functionality and often good performance of several UWB systems have been presented in the past years: starting from ranging experiments [FKML10], through 2D real time localization [SHSM10], up to demonstration of 3D sub-cm accuracy [ZKM$^+$10]. However, a question has to be asked *"how is good performance defined?"*. The pure number is not enough without knowing what the theoretical limit is. For this reason different methods have been developed to assess the theoretical bounds of achievable localization accuracy and precision. However the selected statistical models of errors occurring in the system, which are required to describe those limits, are often based only on assumptions. This in turn results in too optimistic predictions. Additionally, the lack of experimental proof, that a real system can attain the derived bounds, is a common situation [JDW08]. There is also a third research trend, where single system components are optimized to achieve better overall localization performance. Such an approach is however productive only if the influence of single components on the whole system is quantized.

This leads to a third and last fundamental topic of the thesis. Having evaluated the receiver structures based on a self-designed hardware, the entire localization system will be built. The information about characterized components behavior will then be used to derive the theoretical localization error bounds, which in turn will be verified by measurement campaigns. Based on this it will be possible to identify critical system elements and derive the guidelines and methods for localization performance improvement.

The UWB localization topic is also of high interest for the industry. According to [Col08], apart form the research society, the companies such as: Ubi-Sense [Ubi13], Multi-Spectral Solutions (Zebra Enterprise Solutions) [Zeb13], Time Domain [PLU12] (currently PLUS Location Systems), Aether Wire & Location Inc. [Aet10], Mitsubishi Electronic Corp. and Thales [Tha06] are involved in development of commercial UWB localization systems. This is additional motivation for this work, which attempts to give a better insight in the critical aspects of impulse-based Ultra-Wideband localization system design and optimization.

1.2. Goals and outline of the thesis

The goal of the thesis is to develop a concept and realize a precise indoor localization system on the basis of impulse-radio Ultra-Wideband (IR-UWB) technology. During the design, at each stage a detailed analysis of sub-system performance and limitations will be conducted. It must be kept in mind that although the UWB communication aspect lies outside of the scope of this work, UWB systems have inherent data transmission capabilities. The structure of the thesis is schematically depicted in Fig. 1.2.

Evaluation of receiver topologies
As mentioned in the previous section, the selection of optimal receiver architecture is a crucial issue during UWB system design, independent of whether the communication or localization applications are envisioned. One of the goals of this thesis is to develop a comparison criterion for different receiver types and to verify in practice the best performing receiver. For this purpose

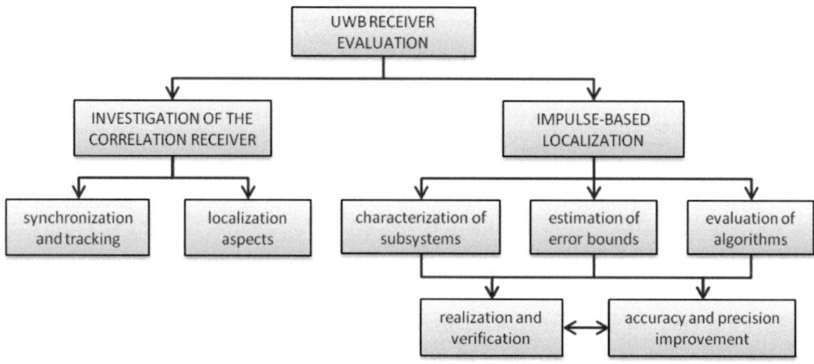

Fig. 1.2.: Representation of the aspects handled in this thesis and their mutual dependency.

a common hardware platform will be developed, that will allow for a fair comparison during measurements. The results will be then compared with theory.

For this purpose, in chapter 2, along side with the fundamental information regarding UWB technology such as regulatory issues and signal shaping, a new comparison metric, called signal-to-threshold ratio (STR), is introduced. Although it is partly similar to the signal-to-noise ratio (SNR), STR is much better suited to compare the impulse-based systems directly in time domain.

Chapter 4 starts with theoretical analysis of IR-UWB transceiver structures encountered in the literature and the selection of candidates for further test. Furthermore, the common system elements are identified, designed and realized in hardware. This creates a generic test platform. The investigation of UWB transceiver structures is done in terms of: achievable transmission range, transmit power efficiency and signal-to-noise ratio in the processed signal. The characteristic of all designed components is in parallel implemented in a system simulator in order to deliver a benchmark for the performed measurements. This part of the investigation is represented by the top block in Fig. 1.2.

Investigation of the IR-UWB correlation receiver
Another aim of the work is to thoroughly investigate the impulse-radio UWB receiver. The emphasis is put on the feasibility of its practical implementation. This includes the hardware realization, as well as the analysis of synchroniza-

tion algorithms and assessment of the system efficiency in terms of wireless connection stability under static conditions.

All investigated aspects of the correlation receiver are described in chapter 5. This includes the analysis of existing synchronization algorithms as well as the essential hardware requirements needed for their implementation. Furthermore, three synchronization algorithms are proposed and parametrized. Their performance, after implementation on an FPGA, is evaluated during measurements. Analog to this, the pulse tracking algorithms are investigated, that allow the stabilization of the wireless transmission. Such a comprehensive study on synchronization of impulse-based correlation receivers, backed-up with measurements, is presented for the first time in the literature.

This part of considerations ends with an experiment to show the potential of the correlation transceiver system in terms of distance measurement applications. This part of the investigation is represented by the three left elements of the diagram in Fig. 1.2.

**Study on realization of an UWB localization system
and its performance bounds**
The most voluminous goal of this thesis is the one devoted to the localization aspects. Here, the aim is to build and thoroughly analyze the performance limits of an IR-UWB localization system. This begins with the analysis, implementation and evaluation of efficient position estimation algorithms. Special emphasis is put on the accuracy of the solution and real-time operation capability. Moreover the investigation of suitable integrity monitoring methods, allowing the detection of incorrect results will be performed. In the next step the RF front-end subsystems and the localization network elements will be characterized to enable the modeling of localization errors and, hence, derivation of the error bounds.

The validation measurements will be performed in small- and large-scale scenarios, equipped with UWB infrastructure to investigate the performance of the designed system in comparison with the derived theoretical limits. Another important aspect that must be taken care of, is the development and evaluation of error compensation methods, to improve the accuracy and precision in the IR-UWB locating systems. This part of the investigation is represented by the right branch of the diagram in Fig. 1.2.

Chapter 3 is devoted to the various aspects of radio localization, explaining the signal measurement techniques, giving an overview of existing solution algorithms and introducing a modified, non-iterative algorithm for calculating the position in systems with relative distance information. After their evaluation the integrity monitoring methods are investigated, that allow the filtering of the incorrect positioning results. Furthermore, the possible localization system layouts, and their influence on the localization performance, are discussed.

In chapter 6, the localization system architecture is proposed and a detailed analysis of crucial system components, such as time measurement unit and antennas, is presented. Based on this, the expression for the theoretical localization precision lower bound is derived.

In chapter 7, the results of extensive tests of the localization demonstrator, built of the elements presented in the previous chapters, are presented and discussed. Additionally, an example measurement performed with a hybrid system, consisting of UWB sensors and an inertial navigation system, is presented and analyzed. The additional advantages of such a solution are the smoother and log-term stable positioning solution.

2. Impulse-radio UWB technology and its characteristic features

To the present day, impulse-based radio technologies are still not as well explored as their narrowband counterparts. In this chapter, the most important, IR-UWB specific aspects are discussed, starting with the choice of the time-domain impulse shape, moving then to regulatory issues, with special emphasis on peak and average power limits. Furthermore, the signal-to-threshold ratio (STR), as a measure of receiver quality, and alternative to the signal-to-noise-ratio (SNR) for impulse signals, is introduced. The chapter is closed with the definitions and differences between three quantities: resolution, accuracy and precision, in the context of localization systems.

2.1. Selection of the impulse shape

The signal is said to be Ultra-Wideband if it occupies a fractional bandwidth of at least 20% or alternatively an absolute bandwidth of 500 MHz. One of the very important aspects in IR-UWB transmission is the spectral efficiency of the signals employed. The radiated short electromagnetic impulse must optimally use the spectral emission mask, defined by the regulations, described in the next section. Hence, the time-domain signal has to be selected in such way, that it will best exploit the available resources in the frequency domain. While considering this technique one always has to keep in mind the feasibility of practical realization and hardware effort.

The pulses in UWB technology, occupying an instantaneous bandwidth of more than 0.5 GHz, have to exhibit a time-domain pulse width in the one-digit nanosecond range or even less. The generation of such pulses would require sampling intervals of few tens of picoseconds [Tim10]. Although digital-to-analog converters (DAC), capable of fulfilling this task are available on the market, this would contradict the idea of cost-efficient transmitter architectures. For this reason, in the majority of applications, the pulse shapes are generated in the analog domain and then forwarded to the antenna.

A number of techniques for the generation of impulses exist, as e.g. damped (attenuated) sine waves or Hermitian pulses [Eis06]. The first, although relatively simple to generate, possess rather poor spectral-domain-efficiency, whereas the last are relatively complicated to generate. On the other hand, signals that are relatively simple to produce with analog electronic circuitry are Gaussian pulses and their derivatives. The Gaussian impulse is among others the pulse shape with the smallest time-bandwidth product. The Gaussian pulse, normalized to the amplitude of 1, can be mathematically described as:

$$p(t) = e^{\left(-\frac{t^2}{2\sigma_p^2}\right)},$$ (2.1)

where σ_p is the pulse standard deviation, controlling its width. The pulse spectrum can be determined with the help of transformation tables:

$$\mathcal{F}(p(t)) = S(f) = \sqrt{2\pi\sigma_p^2} \cdot e^{-2\pi^2\sigma_p^2 f^2}.$$ (2.2)

This spectrum has its maximum at a frequency of $0\,\mathrm{Hz}$, hence it does not fulfill the UWB regulations. Moreover, signals exhibiting a direct current (DC) component cannot be radiated over an antenna. At this point, there are only two options to choose from, if one wants to shift the spectrum upwards in frequency: First, the usage of a conventional mixer and up-conversion of the Gaussian pulse with a carrier frequency. This solution, however, has the drawback of the mixer having only a limited local oscillator (LO) isolation [DDGC11]. The carrier frequency will be radiated as well and most probably violate the allowed spectral emission limit. The second, more elegant option is to build the derivative of the Gaussian pulse $p(t)$. The following relation is valid for the n-th derivative of the Gaussian pulse:

$$\mathcal{F}\left(p^{(n)}(t)\right) = (j2\pi f)^n \cdot S(f),$$ (2.3)

where $(\ldots)^{(n)}$ denotes the n-th derivative. The absolute value of the n-th derivative of the Gaussian pulse spectrum is defined by:

$$\left|S(f)^{(n)}\right| = \left|(j2\pi f)^n \cdot \sqrt{2\pi\sigma_p^2} \cdot e^{-2\pi^2\sigma_p^2 f^2}\right|$$ (2.4)

$$= (2\pi f)^n \cdot \sqrt{2\pi\sigma_p^2} \cdot e^{-2\pi^2\sigma_p^2 f^2}.$$

The center frequency f_M can be found by determining the maximum of the spectrum. For this purpose the absolute spectrum has to be differentiated with regard to frequency:

$$\frac{\partial \left|S(f_M)^{(n)}\right|}{\partial f} \overset{!}{=} 0.$$ (2.5)

With the results from (2.4) the following is obtained:

$$\frac{\partial \left| S(f_M)^{(n)} \right|}{\partial f} = \sqrt{2\pi\sigma_p^2} \cdot (2\pi f_M)^{n-1} \cdot 2\pi n \cdot e^{-2\pi^2\sigma_p^2 f_M^2}$$

$$- \sqrt{2\pi\sigma_p^2} \cdot (2\pi f_M)^n \cdot 4\pi^2\sigma_p^2 f_M \cdot e^{-2\pi^2\sigma_p^2 f_M^2}$$

$$\overset{!}{=} 0. \tag{2.6}$$

Reducing and rewriting of (2.6) results in:

$$\sigma_p = \frac{\sqrt{n}}{2\pi f_M} \tag{2.7}$$

and solved for the center frequency f_M:

$$f_M = \frac{\sqrt{n}}{2\pi\sigma_p}. \tag{2.8}$$

One can recognize, that on the one hand, the center frequency increases proportionally with the square root of the derivative, and on the other hand, that smaller values of the standard deviation σ_p (shorter pulse duration) cause the center frequency to rise as well. In addition, σ_p influences the width of the spectrum. The smaller this parameter is, the broader the spectrum of the impulse becomes. A similar observation was made in [ADA08].

In this work, the adaptation of the pulse shape to fulfill the FCC requirements (see section 2.2) is targeted. For this reason, the 5th derivative of the Gaussian pulse ($n = 5$) has been chosen. The center frequency of the FCC regulation lies at 6.85 GHz and, according to (2.7), the value of the corresponding σ_p is 51.953 ps. The spectra of the 4th, 5th and 6th derivative for the determined value of σ_p are shown in Fig 2.1.

The indoor FCC mask remains unbreached by the Gaussian pulse of the 5th order. The 4th and 6th order pulses violate the mask at 1.61 GHz and 10.6 GHz respectively. The European regulations remain unfulfilled by all pulses.

In the time behavior of the pulses, it can be observed that the absolute value of the signal strength of the original Gaussian pulse at $t = \sigma_p$ is about 60 % of the maximum value. The calculated σ_p of 51.953 ps cannot be considered as the absolute pulse width. Instead $6\sigma_p$ is seen as the absolute pulse width. At a value of $t = 3\sigma_p$ the signal level (rising or falling edge) reduces to 1 % of the maximum. During differentiation the pulses become slightly wider. In the following section, the regulatory issues are analyzed in terms of the allowed pulse amplitude depending on the transmission rate.

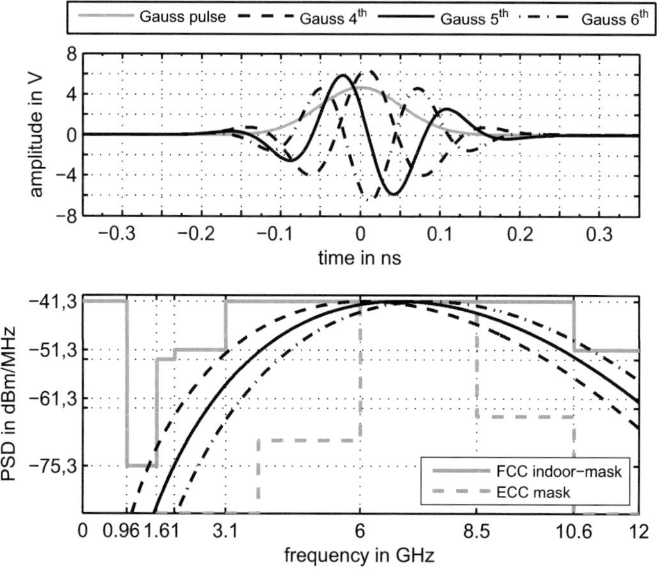

Fig. 2.1.: Time-domain signals (top) and the associated spectra (bottom) of the various Gaussian pulse derivatives. Additionally the spectral emission masks according to FCC and ECC have been overlaid.

2.2. Power emission limits in IR-UWB transmission

The FCC specifies two spectral emission limits for the UWB transmission: one with regard to the average power, and a second one regarding the peak power. In this section, both limits are explained and example calculations are presented for their better understanding.

All specifications and information about power limits in this section refer to EIRP, if not stated otherwise. EIRP is the *equivalent isotropically radiated power* and is defined as follows:

$$\text{EIRP} = P_T \cdot G_T. \tag{2.9}$$

It is the product of the signal power supplied to the antenna (P_T) and the transmit antenna gain (G_T) in a given direction, relative to an isotropic antenna. EIRP limits refer to the highest signal level measured in any direction and at any frequency from the UWB device. The precise test and measurement procedures for different types of UWB devices (e.g. indoor, hand-held, imaging etc.) are specified in the sections 15.31.a and 15.523 of [Fed02]. This section is organized in a way to gain a better understanding of the relations between the average power and peak power emission limits.

2.2.1. Average power limit

As depicted in Fig. 2.1, UWB emission is allowed below 960 MHz and between 3.1 GHz to 10.6 GHz, at a maximum level of −41.3 dBm/MHz. Since the lower range offers much less bandwidth, and is therefore not interesting from the point of view of this work, only the second band will be considered. This part of the regulation refers to the average power emission limit based on RMS (root mean square) measurements performed within a 1 MHz resolution bandwidth (RBW_{avg}). During the measurement a spectrum analyzer with a resolution bandwidth of 1 MHz and activated RMS detector is used. The averaging time has to be set to 1 ms or less.

The average power spectral density (PSD) resulting from the measurement must not exceed −41.3 dBm/MHz. This value corresponds to an average power of −2.55 dBm for a signal occupying 7.5 GHz bandwidth ($P_{\text{rms EIRP}}$), averaged over 1 ms. This value represents the average power of a pulse that perfectly fills the spectral emission mask. However, in practice the spectra of signals do not optimally use this resource allowed in the regulation, and therefore have lower efficiency. The pulse spectral efficiency is defined as [Tim10]:

$$\eta = \frac{\displaystyle\int_{f_L}^{f_H} \text{PSD}_{\text{pulse}}(f) \cdot df}{\displaystyle\int_{f_L}^{f_H} \text{PSD}_{\text{regulation}}(f) \cdot df}, \tag{2.10}$$

where f_L and f_H denote the lower and higher edge of the frequency range of interest. The above expression is an approximation and is valid for signals, which do not exhibit any significant spectral components outside this band. In order to obtain the correct average pulse power, the $P_{\text{rms EIRP}}$ has to be multiplied by the spectral efficiency (2.10) in linear scale; in case when logarithmic scale is used $10 \cdot \log_{10}(\eta)$ has to be added.

In order to relate the average pulse power to either the amplitude or power of a single radiated pulse, the peak power limit will be considered in section 2.2.2.

2.2.2. Peak power limit and maximum pulse amplitude

The limit for the peak level of the emissions is specified as the power contained within a 50 MHz bandwidth (RBW_{peak}), centered around the frequency f_M at which the highest radiated emission occurs. This limit is 0 dBm EIRP. Since standard spectrum analyzers are not equipped with a 50 MHz IF filter, it is acceptable to use other values of resolution bandwidth. This resolution bandwidth, however, must not be lower than 1 MHz nor greater than 50 MHz. If a resolution bandwidth (RBW) other than 50 MHz is applied, the allowed peak power, expressed in dBm is [Fed02]:

$$P_{pk\,RBW} = P_{pk\,50\,MHz} + 20 \cdot \log_{10}(RBW/50\,MHz), \qquad (2.11)$$

where RBW is the applied resolution bandwidth, expressed in megahertz. The peak power measurements have to be conducted with a spectrum analyzer using a peak detector, capturing the signal over a long time period (several minutes, with "maximum hold" function turned on).

In case of a coherent pulse the equation (2.11) is valid for an arbitrary RBW, and hence can be used to calculate the peak power within the entire FCC UWB bandwidth (assuming the pulse exhibits a flat spectrum). As the peak EIRP in 50 MHz bandwidth is limited to 0 dBm, the $P_{pk\,EIRP}$ accumulated within 7.5 GHz equals 43.52 dBm.

One of the most important issues during the design of an impulse-based system is to relate the EIRP to the amplitude of the time domain signal. The calculation of the amplitude corresponding to the peak power limit is described as follows: If the system impedance Z_S of the spectrum analyzer is known, the peak pulse amplitude $A_{pk\,50\,MHz}$ can be analytically determined by solving the below equation:

$$P_{pk\,50\,MHz} = \frac{A_{pk\,50\,MHz}^2}{Z_S}. \qquad (2.12)$$

For a system impedance of 50 Ω and RBW of 50 MHz the allowed maximal peak amplitude is 224 mV. As it is demonstrated in Fig. 2.2, the amplitude scales linearly with the bandwidth. Hence, the maximal pulse amplitude A_{max}, occupying the full 7.5 GHz, is 33.54 V.

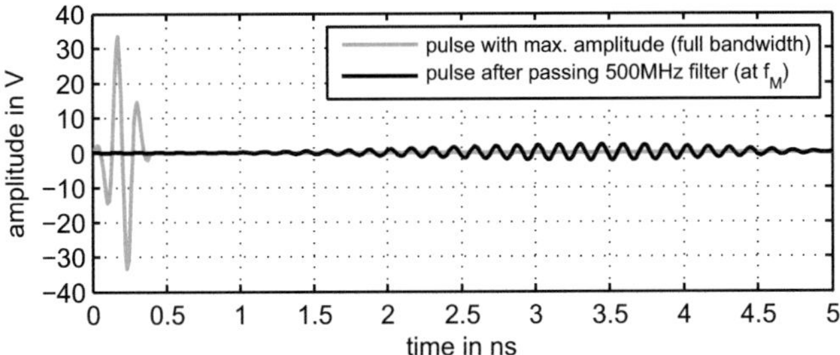

Fig. 2.2.: The UWB pulse with an amplitude of 33.54 V and approx. 7 GHz bandwidth (dashed line) and the same pulse after passing a 500 MHz band-pass filter (continuous line). The amplitude of the filtered pulse (2.24 V) decreases directly proportional with bandwidth.

2.2.3. Turnover frequency and pulse amplitude at an average power limit

The *turnover pulse repetition frequency* (PRF_{tov}) is another parameter is of great interest. If turnover PRF is considered, an impulse UWB device is limited by both the peak power and the average power limit. This can be explained by considering a transmitted train of pulses, where each pulse has an amplitude A_{max} and their spacing equals exactly $T_{tov} = 1/PRF_{tov}$. The T_{tov} describes the separation between pulses, for which the average train power is exactly $P_{rms\,EIRP}$. This value can be determined mathematically, as presented below.

The energy of a single pulse is defined as follows:

$$E_{pulse} = \frac{1}{Z_S} \int_t^{t+\tau_{pulse}} p^2(t)dt = \frac{1}{Z_S} \int_t^{t+\tau_{pulse}} (A \cdot p_{norm}(t))^2 dt, \qquad (2.13)$$

where the τ_{pulse} denotes the pulse duration and $p_{norm}(t)$ is the pulse with normalized amplitude. For the peak power case, the amplitude is equal to A_{max}. Using this relation, the average signal power (power of one period of the above mentioned pulse train) is described with:

$$P_{rms\,pulse} = E_{pulse} \cdot PRF. \qquad (2.14)$$

When using the maximal amplitude in (2.13) and solving the (2.14) for PRF, the turnover frequency can be determined. Another method of calculating the

17

PRF_{tov} may be executed using the expression describing the peak-to-average power ratio (PAR) of a signal. The PAR is defined as follows:

$$\text{PAR} = \frac{P_{\text{pk 50 MHz}}}{P_{\text{rms RBW}_{\text{avg}}}} = \frac{P_{\text{pk 50 MHz}}}{\frac{P_{\text{rms pulse}}}{\eta} \cdot \frac{\text{RBW}_{\text{avg}}}{7.5\,GHz}} . \tag{2.15}$$

The information about PRF_{tov} is included in the average signal power. Rewriting the (2.15) according to the way presented in [Kun04] results in:

$$\text{PRF}_{\text{tov}} = \frac{\text{RBW}^2_{\text{peak}}}{\text{RBW}_{\text{avg}} \cdot 10^{\frac{\text{PAR}}{10}}} . \tag{2.16}$$

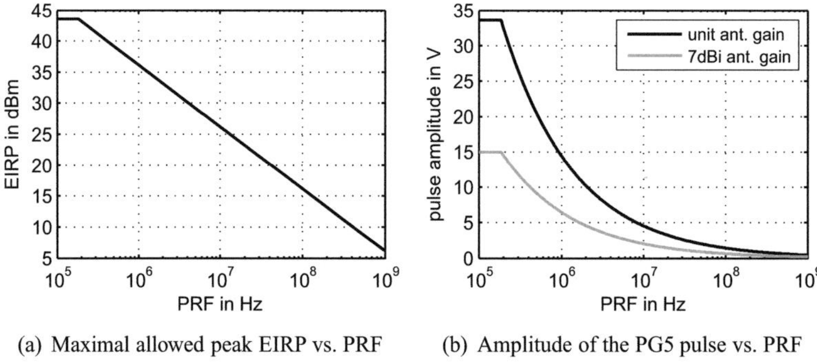

(a) Maximal allowed peak EIRP vs. PRF (b) Amplitude of the PG5 pulse vs. PRF

Fig. 2.3.: Relation between the maximum allowed peak EIRP and the pulse repetition frequency (a). The maximal amplitude of the 5th derivative Gaussian pulse as a function of the pulse repetition frequency is visualized in (b), for the case when an isotropic antenna and antenna with 7 dBi gain are used.

Inserting the values prescribed by the FCC regulation, the PAR equals 41.3 dB and the value of the turnover PRF is 185.328 kHz. An impulse-based UWB device operating below this PRF will be limited by the peak power criterion. For a higher PRF, it will be limited by the average power criterion. This situation is depicted in Fig. 2.3(a), where the maximal allowed EIRP is plotted against the PRF.

In the following consideration an example of a calculation based on the 5th-derivative Gaussian pulse (described in section 2.1) is presented.

The spectral efficiency of this pulse is 51.25%, or equivalently −2.9 dB. In order to calculate the pulse amplitude, corresponding to the peak EIRP, the antenna gain has to be taken into account. As the average power has to remain the same, the pulse amplitude in (2.13) needs to be divided by $\sqrt{G_T}$ and the right side of (2.14) multiplied by G_T. The relation of the maximal 5th-derivative Gaussian pulse amplitude and the pulse repetition frequency is depicted in Fig. 2.3(b). The solid curve shows the run of the amplitude if an antenna with a unit gain is used (0 dBi). The dashed curve describes the maximum allowed amplitude of the pulse at the antenna input with 7 dBi gain.

2.3. Signal-to-noise ratio of pulsed signals

After describing the UWB pulse in its native time domain, a similar description will be done for defining the relationship between pulse and noise amplitudes. In every wireless system, it is desirable to have a high value of the ratio between the useful signal and the noise at the receiver, in order to be able to recover the transmitted signal. In this section, the specific nature of the signal-to-noise ratio (SNR) determination for impulse-based systems will be discussed and a closed formula for its calculation is proposed.

2.3.1. Average signal-to-noise ratio

The classic definition of the signal-to-noise ratio SNR is the relation between the average power of the useful signal in a certain time interval T and the average power of the noise within the same time. Equation (2.17) defines the SNR in a logarithmic scale.

$$\text{SNR}_{\text{dB}} = 10 \cdot \log_{10}\left(\frac{P_{\text{signal}}}{P_{\text{noise}}}\right) = 10 \cdot \log_{10}\left(\frac{A^2_{\text{eff signal}}}{A^2_{\text{eff noise}}}\right) \tag{2.17}$$

The signal power can be calculated according to (2.18) and the AC (alternating current) noise power is the noise voltage variance $\sigma_n^2(t)/Z_S$, as in (2.19).

$$P_{\text{signal}} = \frac{1}{T \cdot Z_S} \int_0^T s_{\text{signal}}^2(t)dt \tag{2.18}$$

$$P_{\text{noise}} = \frac{\sigma_n^2(t)}{Z_S} = \frac{1}{T \cdot Z_S} \int_0^T n^2(t)dt \tag{2.19}$$

Inserting (2.19) into (2.17) results in:

$$\text{SNR}_{\text{dB}} = 10 \cdot \log_{10} \frac{P_{\text{signal}}}{\sigma_n^2(t)/Z_S} = 20 \cdot \log_{10} \frac{A_{\text{eff signal}}}{\sigma_n(t)}. \tag{2.20}$$

The above definition is useful in systems where the signal is constantly present and the average power remains the same independently of the signal section selected for its calculation. This definition can be used in impulse-UWB systems, only in case if at some point in the receiver an integration over entire transmission period is performed. This approach is not suitable for impulse based systems, where no such integration is done (cf. chapter 4) and the signal detection is accomplished basing on the observation of a single pulse. In such case, another definition, based on peak signal and noise power, will be used. Otherwise, the SNR varies with the changing duty cycle of the signal:

$$\text{duty cycle} = \tau_{\text{pulse}} \cdot \text{PRF}_{\text{Tx}} = \frac{\tau_{\text{pulse}}}{t_{\text{Tx-rate}}}, \tag{2.21}$$

where τ_{pulse} denotes the width of the UWB pulse, e.g. the full width at half maximum (FWHM), and PRF_{Tx} stands for the transmission rate of pulses; $t_{\text{Tx-rate}} = 1/\text{PRF}_{\text{Tx}}$. In receivers where no integration over more than one period of the received signal is performed (the case in this thesis), the changing PRF must not influence the SNR.

2.3.2. Signal-to-threshold ratio

In case of receivers which do not integrate the pulse power over a longer time interval (e.g. one Tx period), a different approach has to be introduced, which only involves the peak power (or amplitude) of the pulses:

$$\text{STR}_{\text{dB}} = 10 \cdot \log_{10} \frac{\left(\widehat{A}_{\text{avg}}\right)^2}{\left(3 \cdot \sigma_n(t)\right)^2} = 20 \cdot \log_{10} \frac{\widehat{A}_{\text{avg}}}{3 \cdot \sigma_n(t)}, \tag{2.22}$$

where STR stands for signal-to-threshold ratio and $\widehat{A}_{avg}$ is the mean peak amplitude of the received pulses. The 3σ-limit of the standard deviation is used to account for 99.7% of the normally distributed noise amplitudes (peaks). Both signals are in baseband, measured after low-pass filtering (see also section 2.3.3). The main difference in comparison to (2.20) lies in the fact that the average peak amplitude of pulses is used for the calculation of the STR, instead of the effective value of the signal voltage. However, if the time-domain pulse shape is known the STR can be converted to SNR.

As will be described in section 4.2, broadband correlators are the key elements of UWB receivers. Similarly to the mixers in narrowband systems, they are responsible for the down-conversion of the RF signals to baseband. The equivalent to the LO signal in a conventional mixer is the so-called *template* signal in a correlator. The correlators, just like mixers, have limited isolation of the template signal input to output. An increased noise level can occur, in the case when the template pulse is generated and multiplied with noise instead of with the received pulse. The reason for this is twofold: Firstly the pulses, although small, have a certain cross-correlation with noise (i.e. the absolute value of the correlation coefficient larger than zero). Secondly, cross-talk of the template signal towards the output is present. This combined effect contributes to an effective reduction of the STR. To account for this, equation (2.22) needs to be modified as follows [14]:

$$\text{STR}_{\text{dB}} = 20 \cdot \log_{10}\left(\frac{\widehat{A}_{\text{avg}}}{3 \cdot \sigma_n(t) + A_{\text{xt}}} \right),\qquad(2.23)$$

where A_{xt} represents the average signal amplitude at the output caused by cross-talk and cross-correlation of the template pulse with noise. Using this modified definition, the calculated noise level can be directly adopted as the threshold level for the signal (or bit) detection. The STR offers a basis for a fair performance comparison of different receiver architectures, presented in section 4.4.

2.3.3. Influence of filtering on signal-to-threshold ratio

The considerations regarding the filtering presented in this section are valid for the receiver architectures from section 4.2. This group of receivers will be represented here by the squaring receiver from Fig. 2.4, which additionally is the worst-case example in terms of noise present in the base-band. It consists of a low-noise amplifier, band-pass filter, squaring device and low-pass filter.

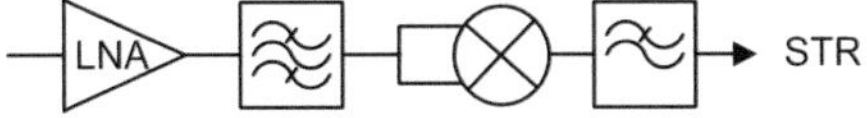

Fig. 2.4.: Receiver model based on squaring device.

The proper selection of both filter bandwidths is a very important aspect, which allows to improve the signal-to-noise ratio. The filtering operation is schematically illustrated in Fig. 2.5(a): in passband (top) and baseband (bottom). First the filtering operation in passband (BPF), after the LNA, is considered. Obviously the selected filter should not be too wide, in order not to pick up excessive out-of-band noise. For a wider filter the signal power remains the same and only noise power increases, hence STR drops. On the other hand, too narrow filter is neither a good selection, as it leads to loss of signal power, hence STR drops as well. Optimal band-pass filter has the same bandwidth as the signal and exhibits steep roll-off.

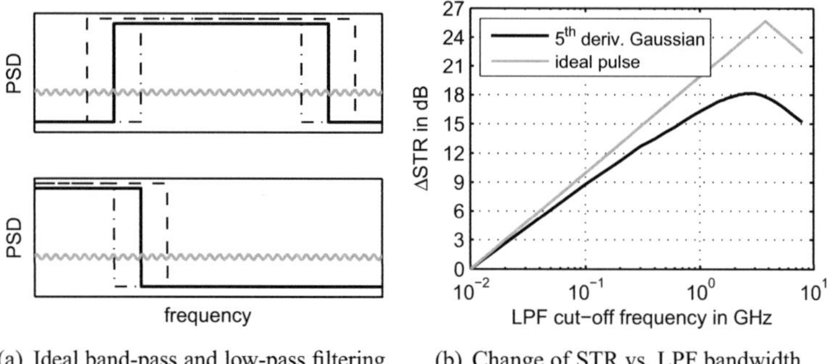

(a) Ideal band-pass and low-pass filtering   (b) Change of STR vs. LPF bandwidth

Fig. 2.5.: Selection of the optimal band-pass and low-pass bandwidths for impulse filtering (a). The solid line represents the signal spectrum, the wavy one noise and the dashed ones the filter characteristics. (b) illustrates the change of the STR as a function of LPF bandwidth for an ideal pulse and the 5th derivative Gaussian pulse.

The squaring operation approximately halves the bandwidth and transfers the energy of the signal and noise into baseband. Following this, the signal has to undergo low-pass filtering (LPF) in order to remove the high frequency components. During squaring, the self-multiplication of noise results in mutual mixing of its frequency components, whereat among others a DC-component is generated. The DC signal however is not taken into account while determining the STR (only AC power is relevant, cf. equation (2.22)) and it can be removed with a high-pass filter (e.g. DC-block). At this point the cut-off frequency of the LPF that will yield the highest possible STR is the parameter of interest.

Optimal low-pass filter bandwidth for an ideal pulse
The amplitude of an ideal UWB pulse (flat spectrum, $\eta = 1$) depends linearly on bandwidth (cf. section 2.2.2), hence when doubling the bandwidth the peak power increases by 6 dB. At the same time the noise amplitude changes with a square-root, causing the increase of variance by 3 dB. Therefore, it is evident that the highest STR is achieved when the LPF bandwidth is equal to the half of the signal bandwidth in passband. The STR will degrade with 3 dB/octave for wider or narrower filter.

This relation is represented by the dashed line in Fig. 2.5(b). Here the change in STR (ΔSTR) is plotted against the corner frequency of the filter. The ΔSTR is determined in relation to the one observed when a LPF with 10 MHz cut-off frequency is used. The shape of the presented characteristic for an ideal pulse remains constant independently of the change in absolute STR.

Optimal filter bandwidth for pulses with lower efficiency
The real pulses, as e.g. the one presented in section 2.1, exhibit spectral efficiency $\eta < 1$. Therefore, the assumption regarding linear scaling of pulse amplitude with bandwidth is valid only within a narrow region close to f_M. This causes that the STR changes at a rate smaller than 3 dB/octave with respect to the filter bandwidth. This characteristic, calculated of a 5th derivative Gaussian pulse, is presented in Fig. 2.5(b) (solid line). It is clearly recognizable that the slope is less steep than for the ideal pulse. Because the spectrum of this particular pulse has a gradual roll-off towards the edges of the UWB band, the curve has less prominent maximum than in the ideal case and the optimal cut-off frequency is also lower. In this case it is 2.8 GHz. For the above calculation a LPF with a roll-off of 55 dB/octave has been used.

In case of pulses that do not exhibit rectangular spectrum, the change of the absolute STR level will affect the run of the above curve. The higher signal amplitude results in shifting the curve towards higher frequencies. On the other hand, the increased noise value will move it to the left. The vertical scaling of the characteristic from Fig. 2.5(b) is dependent on the time domain shape of the pulse and is not directly related to its spectral efficiency.

2.4. Resolution, accuracy and precision

Three terms are often used when describing the localization systems: resolution, accuracy and precision. They are sometimes inadvertently used as synonyms, leading to confusion. They will therefore be explained in this section. Let us start with the term resolution: this is the ability to resolve differences between two or more things. From localization view point, resolution refers to the smallest difference between two positions that can be reliably observed. With a localization system exhibiting high resolution, it is possible to differentiate between two positions lying close to one another.

The difference between the other two terms: accuracy and precision, is explained based on Fig. 2.6. The accuracy of a localization system refers to how much the calculated result differs from the true one, where a reference

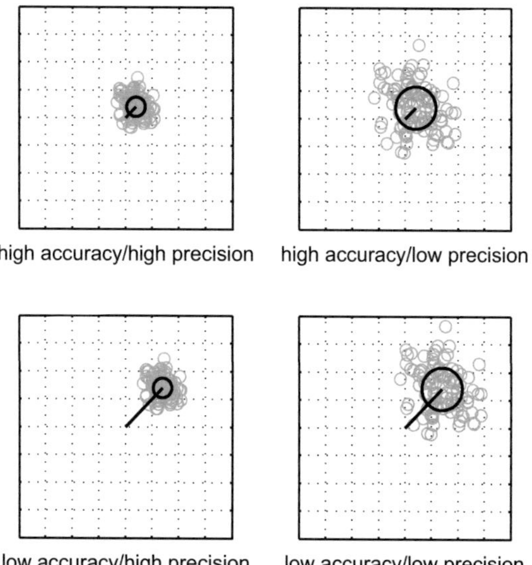

Fig. 2.6.: Graphical representation of the difference between precision (1σ-confidence region marked with black circle) and accuracy (black straight line), based on 100 events (gray circular markers).

measurement is required to determine the true value. The example of a high accuracy is presented in the two upper plots of Fig. 2.6, and is marked as a black, straight line, indicating the difference between the reference (plot center) and the average of all measurements.

Precision, also known as *repeatability*, is the variance in repeated measurements performed in unchanged conditions (e.g. measurement of the same position) [SHH$^+$10]. A reliable measure of precision is the standard deviation (or variance) around the mean value. High precision refers to measurements where the results are focused around a particular point and the variance or standard deviation is small. High precision is shown in the two left plots of Fig. 2.6, and does not automatically imply good accuracy. Low precision indicates a high measurement noise that influences the results. Often, a good method of improving precision is to average the measured parameters before calculating the result. Situations where the precision is high and the accuracy is low often indicate the existence of a systematic error, also called *bias*.

There exist different methods of describing the accuracy:

- accuracy as the difference between the reference position and the average position, calculated from several measurements,

- accuracy as the mean difference between the reference position and all measured ones (in general, this will give the same result as above),

- root-mean-square (RMS) accuracy, or more correctly RMS error, calculated as a square root of the average from the squared differences between reference position and the calculated (from measurements) ones.

By definition, the RMS error is the measure that combines information about the accuracy and the precision. Later in this work, it will be stated which type of accuracy is referred to whenever this term is used.

3. Wireless localization techniques and algorithms

This chapter treats the question of how the position of a user can be determined based on a series of UWB measurements. The main components of the system are the mobile users (MU) that can move within a certain (indoor) scenario, and the access points (AP) that are distributed in this environment. The APs serve as reference nodes and their absolute, or at least relative, positions in the scenario must be known. Due to the reciprocity of the propagation channel, the localization system is reciprocal as well, hence there is no difference (from the physical point of view) whether the AP or MU operate in transmitter or receiver mode. The only difference lies in the practical implementation of both approaches. This will be explained in chapters 6 and 7. The task of calculating the MU position from the measured radio signal parameters is not a trivial problem and requires sophisticated mathematical methods and algorithms.

This chapter is structured as follows: At the beginning, section 3.1 gives an overview of the possible localization methods and explains why the time difference of arrival (TDOA) approach has been selected to develop the localization system in this thesis. In section 3.2, the set of equations, relating the measured signal parameters with the MU position, is defined. The following two sections 3.3 and 3.4 describe the solution of those equations and the complications that can be encountered. Section 3.5 gives an overview of potential error sources which can reduce the quality of the position calculation. For the purpose of identifying the potentially erroneous UWB measurements, the receiver autonomous integrity monitoring (RAIM) algorithms can be applied. Section 3.6 lists the known RAIM algorithms and explains the range-comparison-method implemented in this work. Finally, section 3.8 discusses the problem of the optimal access point placement, based on the so-called dilution of precision (DOP) values.

3.1. Position estimation methods

This section gives an overview of wireless localization methods which can be used in indoor scenarios. The operation principles of those methods are outlined and the individual advantages and disadvantages are discussed in each case.

3.1.1. Received signal strength

The received signal strength (RSS) method estimates the distance between the transmitter and the receiver based on the detected signal power. By employing multiple signal sources (APs) and measuring the corresponding power at the receiver (MU), its position can be calculated by trilateration[4]. In the case of a direct line-of-sight (LOS) propagation, the signal power decreases due to the free space path loss. The UWB technology, thanks to the very short impulses, can help to resolve the direct signal propagation path from echoes. Nevertheless, the received signal power indoors is significantly influenced by the obstacles that the wave encounters on its propagation path. This makes the correct distance calculation complicated, especially under non-line-of-sight (NLOS) conditions. For this reason, often so-called fingerprints are used in practice [BP00, PKC02]. This method is based on gathering the information about the signal power from all APs at discrete positions in the scenario, by means of a measurement campaign. A database, containing the RSS information and the corresponding geometrical coordinates, serves as a look-up table. By the help of this table the most likely position of the MU is estimated, based on signal power readings at a particular location. A more sophisticated version of this approach is described in [GP09], where additional signal parameters and characteristics are evaluated. One of those is the power delay profile (PDP), containing information about the energy content of single echoes. After extracting such characteristics, a more accurate position estimation is possible. However, the creation of the PDP database is significantly more time-consuming. Both methods are not suitable for channels, in which large objects are often displaced, as after every such change a new measurement campaign would be necessary.
A general advantage of the RSS method is that no synchronization is required: neither between AP and MU, nor between APs among each other. However, the determination of the exact signal power in the case of very short UWB

---

[4]lateration - a method of calculating an absolute or relative location of a point by measuring the distances and using the geometry of circles (spheres in 3D) or triangles

pulses is extremely challenging, thus this method is barely applicable for UWB positioning. As demonstrated in table 1.1, the achievable accuracy of RSS, on the order of 1 m, cannot compete with that offered by time-based methods.

3.1.2. Angle of arrival

In the angle of arrival (AOA) method the information about the signal arrival direction is extracted at the receiver using an antenna array. The transmitter position can then be triangulated[5]. If the receiver is capable of measuring both azimuth and elevation angles, then only two receivers are required for a three-dimensional localization. This is graphically represented in Fig. 3.1(a). If one of the coordinates is known a priori (e.g. the height), then even one receiver is sufficient for localization. The further advantage of this concept, similar to the RSS case, is that there is no synchronization required. On the other hand, the impact of the error in angle determination on the localization accuracy will rise linearly with increasing separation between transmitter and receiver. Moreover, the cost of the receiver units is high, as behind each array element a separate UWB receiver is required. A commercially available UWB localization system from Ubisense [Ubi13] is based on a combination of AOA and TDOA (cf. section 3.1.4).

3.1.3. Time of arrival

The time of arrival (TOA) localization approach is based on the absolute signal propagation time measurements between transmitter and receiver. Thanks to the precisely known signal propagation speed in air c_0, it is possible to first calculate the distances, and in the following step to use them for the trilateration of the MU position. This is represented in Fig. 3.1(b). Assuming optimal synchronization between the transmitter and the receivers, three APs are required for a three-dimensional position calculation. For the purpose of signal propagation time calculation, precise information about the transmission and the reception times is required. As in most cases perfect synchronization between MU and APs is not achievable, an additional receiver may be employed to determine the synchronization error. The ideal synchronization between the APs themselves is a prerequisite in this case. Due to the fact that the information regarding the transmission time has to be passed on to

---

[5]angulation - a method of calculating a location of a point by measuring angles between it and
reference points lying on a baseline; no direct distance measurements are required

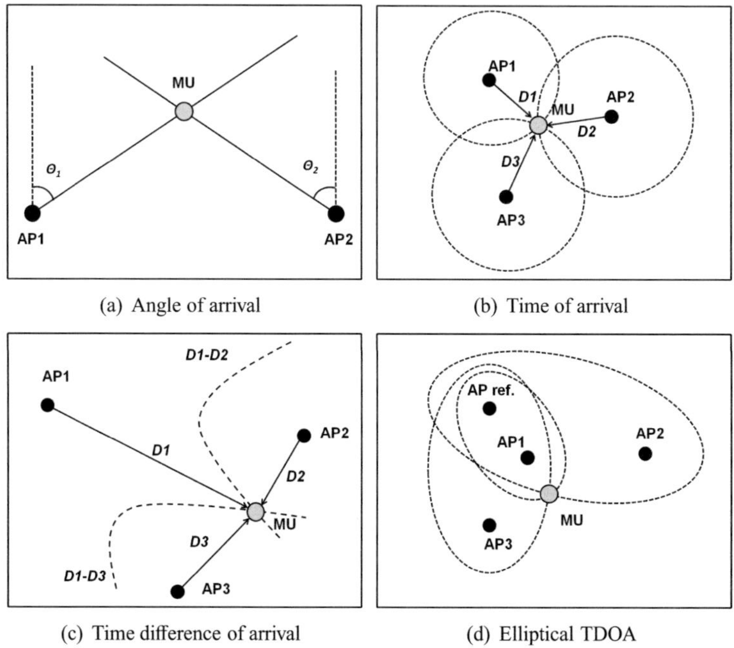

Fig. 3.1.: Graphical representation of localization techniques based on triangulation and trilateration.

the receivers, an additional modulation of the information on the transmitted signal is required. On the one hand, it is possible for the UWB transmitter to simply transmit the actual system time as a data package. Another method would be the use of spreading codes, similar to GPS, to determine the signal time of flight, by using correlators in the digital domain [SGK05].

3.1.4. Time difference of arrival

The time difference of arrival (TDOA) localization method uses the differences in signal propagation times between a MU and multiple APs. The time differences of the signal propagating between MU and two APs trace a curve, for which the distance difference from each point lying on this curve to the AP positions is constant. Such a curve is called a hyperbola and the MU position can be calculated as the intersection of at least two such curves, for which at least three APs are required. This is depicted in Fig.3.1(c).

Thanks to the subtraction of times, the knowledge about the absolute time of signal transmission is not required. This offers the major advantage, compared to TOA, as no internal clock needs to be implemented in the transmitter. In addition, the necessity of synchronizing the MU with the APs becomes obsolete. Only the synchronization among the APs is required. This can be achieved e.g. by connecting the APs either with cables (e.g. LAN) or with conventional wireless communication system (e.g. based on IEEE 802.11 or 802.15.4 standards). The raw data (measured arrival times) are forwarded to a central server, which carries out the position calculation.

The synchronization between the APs can be overcome by the implementation of the round-trip-time method [LS02]. In this case both, the MU and APs, need to be realized as transceivers. The APs, mounted in fixed positions in the scenario, operate as transponders, retransmitting the received signal with a certain delay. After the reception of the signal transmitted by the MU, the AP processes it, and adds information about its coordinates and delay, after which the signal is transmitted back to the mobile device. The MU in turn, receives the UWB impulse after the time corresponding to the round-trip time of flight, plus the processing time in the transponder. Based on this information the MU-AP distance can be calculated. This method is called *two-way-ranging*. In [FKML10], a 1D distance measurement accuracy, using two-way-ranging, of a few centimeters has been demonstrated. The adjacent calculation of the 2D or 3D position, from the MU-AP distances, is based on known absolute ranges and as such is very similar to TOA (cf. Fig. 3.1(b)).

The authors in [ZLGC11] propose another variant of TDOA, which they refer to as *elliptical* localization. In this approach, multiple APs are placed in the scenario, where one of them (preferably mounted centrally) is called the master transmitter. The master AP (mAP) transmits a UWB pulse which is received by the MU and the other APs. The MU then retransmits the pulse which is again recorded by the APs. Each AP can then calculate a TDOA between the 1st pulse from the mAP and the 2nd from the MU. Based on this, the time measurement performed by each AP creates an ellipse, where this particular AP and the mAP are in the focal points. The intersection of the ellipses give then the position of the MU, as depicted in Fig. 3.1(d). Essentially, the two-way-ranging method presented above can be considered as a special case of the elliptical localization, where the two focal points coincide.

Based on the considerations presented so far in this chapter, the TDOA approach, though having the most complicated theoretical description, is the most robust in terms of synchronization. Both the two-way-ranging and the elliptical TDOA localization approaches require precise and stable clocks. The use of repeaters increases the hardware complexity of the system. On the other hand, position calculation based on TOA information is less challenging than in the time difference case. Thus, in this thesis a localization system based on the "pure" TDOA approach has been modeled, built and verified. This comparatively straightforward technique will also allow the direct analysis of single system components influence on the overall performance (cf. section 6).

3.2. TDOA equations

In this section, the system of equations required for the position determination based on measured time differences, is derived. The distance between two points in a Cartesian coordinate system is in general given by the Euclidean norm. Thus, the equation describing the distance between the transmitter and one of the receivers i is the following:

$$c_0 \cdot t_{S,E_i} = \sqrt{(x_S - x_{E_i})^2 + (y_S - y_{E_i})^2 + (z_S - z_{E_i})^2} = \left\| \vec{r}_S - \vec{r}_{E_i} \right\|, \qquad (3.1)$$

where c_0 represents the speed of light in air, t_{S,E_i} for the signal time of flight between transmitter and receiver i. The vectors $\vec{r}$ include the three Cartesian coordinates (x,y,z) of either the transmitter or receiver. This equation is used in the TOA method, because here the distance can be derived directly from the time measurement. Due to the fact that in a TDOA system the differences in the time of flight are measured, the basic equation has the form:

$$c_0 \cdot \Delta t_{ij} = \sqrt{(x_S - x_{E_i})^2 + (y_S - y_{E_i})^2 + (z_S - z_{E_i})^2}$$
$$- \sqrt{(x_S - x_{E_j})^2 + (y_S - y_{E_j})^2 + (z_S - z_{E_j})^2} \qquad (3.2)$$
$$= \left\| \vec{r}_S - \vec{r}_{E_i} \right\| - \left\| \vec{r}_S - \vec{r}_{E_j} \right\|, \quad i,j = 1,\dots,N \text{ and } i \neq j.$$

Δt_{ij} is the propagation time difference the between locations where APs i and j are placed and the position of the MU:

$$\Delta t_{ij} = t_{S,E_i} - t_{S,E_j}. \qquad (3.3)$$

For simplification, from now on, instead of writing t_{S,E_i} the short form t_i will be used, and instead of $c_0 \cdot \Delta t_{ij}$ the distance difference Δd_{ij} will be used. Equation (3.2) describes a three-dimensional hyperbola in space, a so called

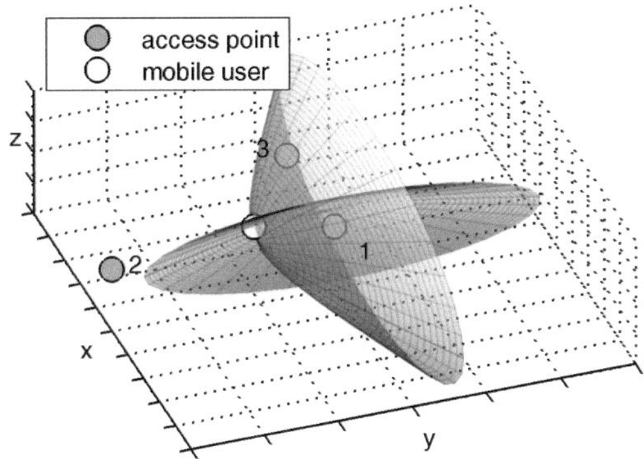

Fig. 3.2.: Hyperbolic trilateration in the TDOA localization method.

hyperboloid. In words, the hyperbola is defined as a set of points, where the absolute value of the distance differences to the two focal points is constant and equal to twice the distance between its two vertices. That is why the TDOA method is sometimes referred to as *hyperbolic trilateration*. In Fig. 3.2, the two hyperboloids are depicted, representing the surfaces of the constant distance differences between the APs 1 and 2 as well as 1 and 3. This results in an intersecting set (hyperbola), where the transmitter position can lie.

For N receivers, a total of $\binom{N}{2}$ time differences can be built. This number corresponds to all possible double-combinations of receivers. In this set, there are however only $N-1$ equations which are linearly independent. As an example, the time difference t_{23} can be written as:

$$t_{23} = t_2 - t_3 = (t_1 - t_3) - (t_1 - t_2) = t_{13} - t_{12}. \tag{3.4}$$

Therefore, for N receivers a maximum number of $N-1$ linearly independent equations can be established. The AP, with respect to which all the differences are calculated, will be referred to as the *base receiver, (BR)*. In order to determine the three unknown transmitter coordinates, four receivers are required. This configuration is represented in Fig. 3.3.

The resulting set of equations is nonlinear, and as such has a nontrivial solution. The unknown values are the coordinates of the transmitter and are summarized in $\vec{r}_S$. Equation (3.2) can be written in a more compact way as:

$$\Delta d_{1j} = h(\vec{r}_S), \quad j = 2, \ldots, N, \tag{3.5}$$

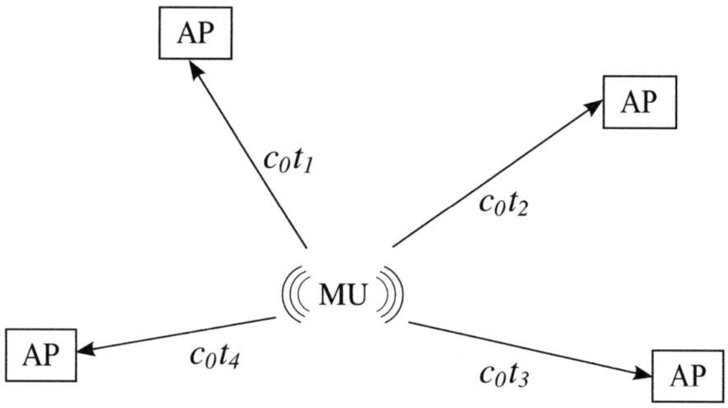

Fig. 3.3.: Signal time of flight between transmitter and different receivers.

where the function h represents the nonlinear projection of the transmitter position on the distance difference Δd_{ij}. In this equation, the BR has already been introduced and marked with the index 1. For the purpose of a more compact notation, the time differences shall be summarized in a vector $\vec{\rho}$:

$$\vec{\rho} = \begin{bmatrix} \Delta d_{12} & \Delta d_{13} & \Delta d_{14} & \cdots \end{bmatrix}^{\mathrm{T}}. \tag{3.6}$$

The indexing of the vector $\vec{\rho}$ refers to the receiver number involved in building the difference to the BR. E.g. ρ_{12} is the first entry of the vector $\vec{\rho}$, which corresponds to the distance difference Δd_{12}. Thus the system of equations of the TDOA method has the following form:

$$\begin{aligned}
\Delta d_{12} &= \left\| \vec{r}_S - \vec{r}_{E_1} \right\| - \left\| \vec{r}_S - \vec{r}_{E_2} \right\| \\
\Delta d_{13} &= \left\| \vec{r}_S - \vec{r}_{E_1} \right\| - \left\| \vec{r}_S - \vec{r}_{E_3} \right\| \\
&\;\;\vdots \qquad\qquad \vdots \\
\Delta d_{1j} &= \left\| \vec{r}_S - \vec{r}_{E_1} \right\| - \left\| \vec{r}_S - \vec{r}_{E_j} \right\|.
\end{aligned} \tag{3.7}$$

The residual f_{1j} at the position $\vec{r}_S$ is the difference between the measured distance difference $\tilde{\rho}_{1j}$ and the value of the function from (3.2) at the point $\vec{r}_S$:

$$f_{1j}(\vec{r}_S) = \tilde{\rho}_{1j} - \rho_{1j} = \tilde{\rho}_{1j} - \left(\left\| \vec{r}_S - \vec{r}_{E_1} \right\| - \left\| \vec{r}_S - \vec{r}_{E_j} \right\| \right). \tag{3.8}$$

In the case of a perfect measurement, the estimated transmitter position $\vec{r}_S$ is equal to the true position and all residuals are equal to zero. If the estimated position differs from the true one, the measurements and the function value do not match anymore, whereby the residuals grow. For N receivers, $N-1$

residues can be calculated. Generally, the sum of the squared residuals is used as quality criterion to find the optimum solution:

$$F(\vec{r}_S) = \sum_{j=2}^{N} f_{1j}^2(\vec{r}_S).$$ (3.9)

The function F is often referred to as the observation function of the error function. This error function is to be minimized during the position calculation:

$$\hat{\vec{r}}_S = \underset{x,y,z}{\mathrm{argmin}}\, F(\vec{r}_S).$$ (3.10)

In order to achieve this, a diverse nonlinear optimization algorithms can be used. Some of them will be explained in section 3.4. However, in order to allow for a proper functionality of the localization algorithms, the correlation of the time measurements needs to be treated first. This issue will be covered in the next section.

3.3. Decorrelation of the measured time values

The work on deccorelation of TDOA measurements presented in this section has been conducted in cooperation with the author of [35] and published in [18].
The necessity of decorrelating the measured time values may not be clear at the first glance. In the previous section, the TDOA equations were derived without the consideration of any uncertainties. However, real measurements are always afflicted with errors (cf. section 3.5). Those can be accounted for by introducing additional noise terms in (3.2). Each time measurement is afflicted by an uncertainty:

$$c_0 \cdot \tilde{t}_{S,Ei} = \left\| \vec{r}_S - \vec{r}_{E_i} \right\| + w_i,$$ (3.11)

where w_i denotes noise term on the i-th AP. Thus, during the calculation of differences, the correlated noise components occur:

$$\tilde{\rho}_{1j} = \Delta \tilde{d}_{1j} = \left\| \vec{r}_S - \vec{r}_{E_1} \right\| - \left\| \vec{r}_S - \vec{r}_{E_j} \right\| + w_1 - w_j \quad, j = 2,\ldots,N.$$ (3.12)

From this, it is obvious that the measurement noise from the BR is present in all other measurements. This happens regardless of which AP is chosen as a reference and results in a correlation of noise terms. In the least squares method, used in section 3.4, as well as for the calculation of the DOP values in section 3.8, uncorrelated measurement values are required. In order to find the correlation terms, the covariance matrix of the measurement vector $\tilde{\vec{\rho}}$ needs to be derived.

3.3.1. Derivation of the distance-difference-vector covariance matrix

The covariance matrix of a vector specifies all pair-wise covariations of all vector elements. The diagonal entries of the covariance matrix are the auto-covariances - the covariances of the same measurement values. The cross-covariances can be found in the secondary diagonal and they describe the dependency of one measurement value on the others. In the following, the covariance matrix of the measurement vector from (3.12) will be derived. The covariance matrix is defined as:

$$cov\{\tilde{\vec{\rho}}\} = E\left\{\left[\tilde{\vec{\rho}} - E\{\tilde{\vec{\rho}}\}\right] \cdot \left[\tilde{\vec{\rho}} - E\{\tilde{\vec{\rho}}\}\right]^{\mathrm{T}}\right\}. \tag{3.13}$$

In case when the errors are assumed to have zero-mean, the expected value of the measurement vector entries $\tilde{\vec{\rho}}$ are:

$$E\{\tilde{\rho}_j\} = E\left\{\left\|\vec{r}_S - \vec{r}_{E_1}\right\| - \left\|\vec{r}_S - \vec{r}_{E_j}\right\| + w_1 - w_j\right\} \tag{3.14}$$

$$= E\left\{\left\|\vec{r}_S - \vec{r}_{E_1}\right\| - \left\|\vec{r}_S - \vec{r}_{E_j}\right\|\right\} + E\{w_1\} - E\{w_j\}$$

$$= \left\|\vec{r}_S - \vec{r}_{E_1}\right\| - \left\|\vec{r}_S - \vec{r}_{E_j}\right\|.$$

After inserting this into (3.13), it simplifies to:

$$cov\{\tilde{\vec{\rho}}\} = E\left\{\begin{bmatrix} w_1 - w_2 \\ w_1 - w_3 \\ \vdots \end{bmatrix} \cdot \begin{bmatrix} w_1 - w_2 & w_1 - w_3 & \cdots \end{bmatrix}\right\} \tag{3.15}$$

$$= E\left\{\begin{bmatrix} (w_1 - w_2)^2 & (w_1 - w_2)(w_1 - w_3) & \cdots \\ (w_1 - w_3)(w_1 - w_2) & (w_1 - w_3)^2 & \cdots \\ \vdots & \vdots & \vdots \end{bmatrix}\right\}.$$

After the expansion of the diagonal entries, and under assumption that the noise therms w_i have normal distribution (with identical standard deviations

$\sigma_{d,i} = \sigma_d$) and are uncorrelated (cf. section 6.2), the diagonal entries of the covariance matrix simplify as in (3.16).

$$E\left\{(w_1 - w_j)^2\right\} = E\left\{w_1^2\right\} - \underbrace{E\left\{2w_1 w_j\right\}}_{=0} + E\left\{w_j^2\right\} = 2\sigma_d^2 \qquad (3.16)$$

In a system where all receivers have the same architecture, i.e. consist of identical components and the time measurements are performed by the same hardware, such equal noise standard deviation for all APs can be safely assumed. In this case, the secondary diagonal simplifies to:

$$E\left\{(w_1 - w_i)(w_1 - w_j)\right\} = E\left\{w_1^2\right\} - \underbrace{E\left\{w_1 w_i\right\}}_{=0} - \underbrace{E\left\{w_1 w_j\right\}}_{=0} + \underbrace{E\left\{w_i w_j\right\}}_{=0} = \sigma_d^2.$$

$$(3.17)$$

Using the result of (3.16) and (3.17), the complete covariance matrix can be written as:

$$\mathbf{C} = cov\left\{\vec{\tilde{\rho}}\right\} = \begin{bmatrix} 2\sigma_d^2 & \sigma_d^2 & \sigma_d^2 & \cdots \\ \sigma_d^2 & 2\sigma_d^2 & \sigma_d^2 & \cdots \\ \vdots & \vdots & \vdots & \vdots \end{bmatrix}. \qquad (3.18)$$

The diagonal entries contain the double standard deviation of a time measurement and in the secondary diagonal the single standard deviation is found. The latter entries are the reason why the measurement values need to be decorrelated. Should this fact be ignored, then for the same measurement a different positioning solution will be obtained, every time the BR is changed. The decorrelation can be achieved with the Cholesky decomposition of the covariance matrix [MÖ8], and will be explained in the next section.

3.3.2. Cholesky decomposition of the covariance matrix

The Cholesky decomposition is an instrument of numerical mathematics which is used for the decomposition of a symmetrical, positively defined matrix into a diagonal matrix $\mathbf{D}$ and a triangular matrix. For the lower triangular matrix (a.k.a. left triangular matrix) $\mathbf{L}$ and for the upper triangular matrix (a.k.a. right triangular matrix) $\mathbf{U}$ will be used. Because of the matrix names, this decomposition can also be found in the literature under the names LD- or UD-decomposition. The covariance matrix from (3.18) is clearly symmetrical and all elements are positive, thus the LD-decomposition, according to the formula:

$$\mathbf{C} = \mathbf{LDL}^{\mathrm{T}} \qquad (3.19)$$

can be performed. The decorrelated covariance matrix $\mathbf{D}$ results from the linear transformation with the matrix $\mathbf{L}^{-1}$:

$$\mathbf{D} = \mathbf{L}^{-1}\mathbf{C}\mathbf{L}^{-1,T}. \tag{3.20}$$

Analogous to this, the measurement vector $\tilde{\vec{\rho}}$ has to be transformed with $\mathbf{L}^{-1}$ according to:

$$\tilde{\vec{\rho}}' = \mathbf{L}^{-1}\tilde{\vec{\rho}}. \tag{3.21}$$

Further information on the decorrelation of the measurement values can be found in [GA08]. In the following section 3.4, the measurement equation (3.2) will be linearized, which results in a matrix $\mathbf{H}$, that also needs to be transformed with $\mathbf{L}^{-1}$. The resulting decorrelated covariance matrix $\mathbf{D}$ is a diagonal matrix, whose entries are not identical. This has to be considered in the further procedure.

The diagonal entries of the k-th line and column of the $\mathbf{D}$ matrix can be written as:

$$\mathbf{D}_{kk} = \left(1 + \frac{1}{k}\right) \cdot \sigma_d^2 \tag{3.22}$$

and the inverse of the lower triangular matrix $\mathbf{L}$ is:

$$\mathbf{L}^{-1} = \begin{bmatrix} 1 & 0 & 0 & 0 & \cdots \\ \frac{1}{2} & 1 & 0 & 0 & \cdots \\ \frac{1}{2} & \frac{1}{3} & 1 & 0 & \cdots \\ \vdots & \vdots & \vdots & \vdots & \vdots \end{bmatrix}. \tag{3.23}$$

In this way, complex calculations in every step, especially when inverting the $\mathbf{L}$ matrix, can be avoided. After decorrelating the measured TDOAs, in the next section the methods of calculating the MU position based on those values will be discussed.

3.4. Estimating the location from time of measurements

In order to calculate the MU position from the pulse time of flight measurements, the error function (3.9), defined in section 3.2, needs to be minimized. In theory, this can be done by investigating the value of this function for every input argument. As a result, one would obtain an error landscape, where the position of its global minimum would correspond to the MU position. However, this approach is highly time consuming (especially for large scenarios and high result accuracy). Therefore, a number of dedicated algorithms for solving such nonlinear problems have been created. They all follow the same idea: first, a rough estimation of the solution is performed, which then can be interpreted as a starting point in the error landscape. From this point, a descent direction in the error landscape is calculated. This direction is calculated in a way that the reduction of the error function value becomes most likely. Subsequently, a first step with a certain step width, from the starting point in the descent direction is done. A new point is then reached and serves as a new starting point. The same procedure is then applied iteratively until a stop criterion is met. Often, either the change in the error function or the change in the calculated position between two steps are used as a stop criterion. If this value is small enough, it can be assumed that a global (or only local) minimum has been reached.

In the following sections, the diverse techniques for solving the nonlinear system of equations will be presented. Besides the Gauss-Newton and the Levenberg-Marquardt methods, there exist very sophisticated ones such as Line-Search- and Trust-Region-algorithms. To avoid illogical solutions of MU position (e.g. outside the building) the auxiliary conditions can be incorporated in the solving algorithm. The Bancroft-algorithm has originally been used for the non-iterative solving of the GPS equations, obtained from TOA measurements. In section 3.4.6, a modified version of this algorithm is introduced, which allows for solving the TDOA equations. The performance of all implemented algorithms are finally compared with respect to their accuracy and computational effort.

3.4.1. Gauss-Newton-algorithm

The Gauss-Newton-algorithm (GN) can be derived in the easiest way if the system of equations (3.7) is linearized around the starting point $\vec{r}_{S,0}$ and afterwards solved using a least squares estimator. The linearization of the equations can be achieved by approximating them with a 1st-order Taylor series. The starting vector results from the start values for the x-, y- and z-coordinates of the MU:

$$\vec{r}_{S,0} = \begin{bmatrix} x_{S,0} & y_{S,0} & z_{S,0} \end{bmatrix}^{\text{T}}. \tag{3.24}$$

The Taylor series approximation of the 1st-order od (3.2) is:

$$\rho_{1j} = \underbrace{\rho_{1j}\big|_{\vec{r}_S=\vec{r}_{S,0}}}_{\rho_{1j,0}} + (x_S - x_{S,0}) \cdot \underbrace{\frac{\partial \rho_{1j}}{\partial x_S}\bigg|_{\vec{r}_S=\vec{r}_{S,0}}}_{h_{1j,x}} + (y_S - y_{S,0}) \cdot \underbrace{\frac{\partial \rho_{1j}}{\partial y_S}\bigg|_{\vec{r}_S=\vec{r}_{S,0}}}_{h_{1j,y}}$$

$$+ (z_S - z_{S,0}) \cdot \underbrace{\frac{\partial \rho_{1j}}{\partial z_S}\bigg|_{\vec{r}_S=\vec{r}_{S,0}}}_{h_{1j,z}} \quad , \quad j = 2, \ldots, N. \tag{3.25}$$

If the first term on the right side $\rho_{1j,0}$ is moved to the left, then (3.25) can be written as a linear system of equations, for N APs:

$$\underbrace{\begin{bmatrix} \rho_{12} - \rho_{12,0} \\ \rho_{13} - \rho_{13,0} \\ \vdots \\ \rho_{1N} - \rho_{1N,0} \end{bmatrix}}_{\text{vector of residues } \Delta\vec{\rho}} = \underbrace{\begin{bmatrix} h_{12,x} & h_{12,y} & h_{12,z} \\ h_{13,x} & h_{13,y} & h_{13,z} \\ \vdots & \vdots & \vdots \\ h_{1N,x} & h_{1N,y} & h_{1N,z} \end{bmatrix}}_{\text{measurement matrix } \mathbf{H}} \cdot \underbrace{\begin{bmatrix} x_S - x_{S,0} \\ y_S - y_{S,0} \\ z_S - z_{S,0} \end{bmatrix}}_{\text{solution vector } \Delta\vec{r}} . \tag{3.26}$$

$\Delta\vec{\rho}$ is the vector of residues at the position $\vec{r}_{S,0}$ (cf. (3.8)). The matrix $\mathbf{H}$ (mapping matrix) is the linear projection of the vector $\Delta\vec{r}$ on the distance differences. In other words, it relates the measured time differences to the differences between the starting and the calculated position. This matrix consists of the differentiation results of the nonlinear function $h(\vec{r}_S)$ (from (3.5)) with respect to $\vec{r}_S$. It is thus identical with the Jacobi matrix of the function h. The derivatives of the function h with respect to the variables (x, y, z) result in:

$$\frac{\partial \rho_{1j}}{\partial x_S} = \frac{x_S - x_{E_1}}{\left\|\vec{r}_S - \vec{r}_{E_1}\right\|} - \frac{x_S - x_{E_j}}{\left\|\vec{r}_S - \vec{r}_{E_j}\right\|},$$

$$\frac{\partial \rho_{1j}}{\partial y_S} = \frac{y_S - x_{E_1}}{\left\|\vec{r}_S - \vec{r}_{E_1}\right\|} - \frac{y_S - y_{E_j}}{\left\|\vec{r}_S - \vec{r}_{E_j}\right\|}, \tag{3.27}$$

$$\frac{\partial \rho_{1j}}{\partial z_S} = \frac{z_S - x_{E_1}}{\left\|\vec{r}_S - \vec{r}_{E_1}\right\|} - \frac{z_S - z_{E_j}}{\left\|\vec{r}_S - \vec{r}_{E_j}\right\|}.$$

The system of equations (3.26) can be solved using the least squares approach, which minimizes the sum of error squares. The condition for the unbiasedness of the estimation (in average the true value) and efficiency (minimal variance of the estimation) of the least squares method is, that the measurement values must not be correlated and have an identical variance [DR08]. The decorrelation of the values can be achieved by performing the LDL-decomposition described in section 3.3. For this purpose, the measurement matrix $\mathbf{H}$ and the measurement vector $\vec{\rho}$ need to be transformed with the inverse of the lower triangular matrix $\mathbf{L}$:

$$\Delta\vec{\rho}' = \mathbf{L}^{-1} \cdot \Delta\vec{\rho}, \tag{3.28}$$

$$\mathbf{H}' = \mathbf{L}^{-1} \cdot \mathbf{H}. \tag{3.29}$$

Due to the fact that after the above transformation the measurement values do not exhibit identical variances, the conventional least squares estimator is indeed unbiased, however not efficient. To ensure its efficiency, a weighted least squares approach (or generalized least squares) is required [Ham94]. This estimator weights the residues according to the variance of the measurement values. Doubtful measurements are considered with less weight than those with smaller variance. The weights g result from the reciprocal diagonal entries of the decorrelated covariance matrix $\mathbf{D}$.

$$F(\vec{r}_S) = \sum_{j=2}^{N} g_{1j} \cdot f_{1j}^2(\vec{r}_S) \tag{3.30}$$

The solution of the equation set (3.26) with the weighted least squares estimator is:

$$\Delta\vec{r} = \left(\mathbf{H}'^{\mathrm{T}}\mathbf{D}^{-1}\mathbf{H}'\right)^{-1} \mathbf{H}'^{\mathrm{T}}\mathbf{D}^{-1} \cdot \Delta\vec{\rho}, \tag{3.31}$$

where the matrix $\mathbf{D}$ is the decorrelated covariance matrix from section 3.3.2. The positioning solution $\vec{r}_S$ results from the step of the length 1 in the direction $\Delta\vec{r}_S$:

$$\vec{r}_S = \vec{r}_{S,0} + \Delta\vec{r}. \tag{3.32}$$

The localization solution, however, is accurate only if the errors caused by the linearization in (3.25) remain small. The further away the starting position $\vec{r}_{S,0}$ is from the true position, the larger the error. For large distances, the linearization errors increase very fast and the calculated solution will not be optimal. This error can be improved iteratively, by repeated application of the procedure described above. For this purpose, the equations are repeatedly linearized, new residue vector calculated, solved with (3.31) and the position is updated with (3.32). This is repeated until the sum of squared errors barely changes between iterations.

The choice of the starting position significantly influences the speed of the algorithm. The further the initial guess is from the true position, the more iterations are required. Furthermore, for starting positions, which are far from the true position, the risk of slipping into a local minimum of the function increases. Besides a global minimum, the error function often has local minima. If the algorithm encounters such a local minimum in the error landscape, it will be stuck at that point, because the derivative will approach zero.

Moreover, the convergence properties of the algorithm can cause problems. If the system of equations is poorly conditioned, large step lengths $\Delta\vec{r}$ can occur. The linear approximation, however, is only valid in the proximity of the starting point. This circumstance can cause the lack of convergence of this method [HB06]. The Levenberg-Marquardt algorithm extends the GN in order to better deal with this particular difficulty. It will be described in the next section. To improve the convergence properties of the GN algorithm, the Line-Search strategies can be applied. They will be described in section 3.4.3.

3.4.2. Levenberg-Marquardt-algorithm

The Levenberg-Marquardt (LM) algorithm is a small but significant extension of the GN algorithm. By introducing the so-called attenuation parameter λ, the descent direction can now be controlled in every iteration step.

$$\Delta\vec{r} = \left(\mathbf{H}'^{\mathrm{T}}\mathbf{D}^{-1}\mathbf{H}' + \lambda\,\mathrm{diag}\left(\mathbf{H}'^{\mathrm{T}}\mathbf{H}'\right)\right)^{-1}\mathbf{H}'^{\mathrm{T}}\mathbf{D}^{-1} \cdot \Delta\vec{\rho} \qquad (3.33)$$

The control of the attenuation coefficient λ is performed by an evaluation of the residues. If the norm of the residues from (3.30) became larger in the current step, this means the approximation of the nonlinear function is worse at this position. In that case, λ is increased by a certain factor and the evaluation at this position is re-performed. In practice, λ is increased by one order of magnitude. Should the residue norm decrease in the current step, the approximation is good and the λ is reduced by one order of magnitude for the next step. In general, the initial value of λ is generated heuristically. In this work the starting value has been chosen to $\lambda = 0.01$.

As already mentioned, this procedure improves the convergence performance of the algorithm. Through this additional parameter λ the reduction of the residue norm can be enforced in every step. This is achieved at a cost of a slightly higher computational effort, as under certain circumstances equation (3.33) will need to be evaluated multiple times within one step. It is possible to implement the Levenberg-Marquardt algorithm as a Trust-Region algorithm. This class of nonlinear optimization algorithms will be explained in the next section.

3.4.3. Trust-Region- and Line-Search-algorithms

The Trust-Region (TRR) algorithm draws the conclusions from the fact that the linear approximation of the function is only valid in a certain region around the linearization point. The trust region states, in which area the linear approximation can be considered as sufficiently good. In the case when the reduction of the error function is not sufficient or if the error function value increases, the evaluation is performed again, this time however with a reduced trust region. This is repeated until a satisfactory reduction of the error function is achieved. Should this reduction be large enough within one step, then the linearization will be performed in the new position with an increased trust region. With this method, the convergence problems can be effectively eliminated. More information on this topic can be found in [CCT00].

Whereas in the TR-algorithm the step width is first estimated by the trust region and then the descent direction is defined, the Line-Search method operates the other way round: first, the appropriate descent direction is chosen and then the step length is optimized. This can be achieved in (3.32) by introducing a step length α:

$$\vec{r}_S = \vec{r}_{S,0} + \alpha \cdot \Delta \vec{r}. \tag{3.34}$$

The optimization of the parameter α is a one-dimensional optimization problem, which explains the origin of the name Line-Search. More information on this can be found in [NW05, Chr09].

3.4.4. Consideration of auxiliary conditions

In the case of imperfect or even faulty measurements, the algorithms presented until now can produce irrational solutions, as e.g. those lying outside of the specified room or building. In order to avoid such a situation, the auxiliary conditions need to be considered while solving the system of equations. Those can be formulated as inequations with regard to the MU position $\vec{r}_S$:

$$\vec{r}_{S,L} \leq \vec{r}_S \leq \vec{r}_{S,U}, \tag{3.35}$$

$\vec{r}_{S,L}$ being the lower and $\vec{r}_{S,U}$ the upper limit. The basic principle of the non-linear bounded optimization is to combine the auxiliary conditions with the error function F and to transform it to an unbounded minimization problem with a modified error function. The auxiliary condition can be incorporated by using *barrier functions, penalty functions* or *interior functions*. These terms, added to the error function, need to ensure that it will achieve very high values outside of the limits defined by the inequality (3.35). As a result, the minimum of the new error function will certainly be within the specified borders. The unbounded minimization problem can then be solved using well-established methods.

An algorithm that yields very good results in this work is the Interior-Point (IP) algorithm. The original error function F is here extended by the *interior function* (e.g. logarithmic function):

$$F'(x,\mu) = F(x) - \mu \cdot \underbrace{\sum_{i=1}^{m} \log(c_i(x))}_{I_{log}(x)}, \quad \mu \geq 0. \tag{3.36}$$

The c_i includes the auxiliary conditions that need to be converted into a form $c_i \geq 0$ before. As consequence, the value of the interior function $I_{log}(x)$ for $c_i \to 0$ increases toward infinity. More information on this can be found in [CBG99, FGW02].

3.4.5. Localization in two dimensions

The localization solution in 2D in this case without considering the height can be interesting for two reasons. First, for the calculation of the two-dimensional solution one AP less is required, as the number of degrees of freedom reduces by one. Hence, even in the regions where as few as 3 APs are available, a 2D localization can still be performed. Second, in the case when the UWB- is merged with an integrated pedestrian navigation (IPNS) system (see section 7.3), the task of height estimation can be performed by the integrated barometer. Additionally, due to the often reduced height accuracy of the UWB system, even in the presence of a sufficient number of APs (cf. section 3.8) it can sometimes be reasonable to use only the horizontal information and obtain the vertical information by other means.

All the algorithms used before for the 3D positioning can as well be employed to perform the 2D localization. In this case, only the minimization problem changes slightly: a constant height $z_{S,c}$ of the MU has to be assumed. The new position vector of the MU now has the form $\vec{r}_S = [x_S \; y_S \; z_{S,c}]^T$, where only x_S and y_S are the unknown values. The residue vector can be calculated with (3.8). As $z_{S,c}$ is now constant, the derivation with respect to this coordinate disappears from (3.26), resulting in a $(N \times 2)$-matrix. The other derivatives appearing in this matrix can just as before be calculated with (3.27).

3.4.6. Bancroft-algorithm and its TDOA modification

Stephen Bancroft proposed an algorithm for the discrete solution of a systems of GPS equations [Ban85]. However, because GPS operates based on the TOA principle (see section 3.1.3), the equations have a different form from those of the TDOA system. By reformulating the system of equations from (3.2), it is possible to use this approach to solve the TDOA case. In this section, the modified Bancroft (mBA) algorithm is described. This concept has been developed in cooperation with co-authors of [4] and published in [5].

The first step is to introduce a new variable b, describing the absolute distance between the MU and the reference AP:

$$b = -\left\| \vec{r}_S - \vec{r}_{E1} \right\|. \tag{3.37}$$

In such way, a new system of equations can be created, which contains one additional variable b, however, in return also the additional equation (3.37). The resulting equations can be written as:

$$\begin{aligned}
0 &= \left\| \vec{r}_S - \vec{r}_{E_1} \right\| + b \\
\Delta d_{21} &= \left\| \vec{r}_S - \vec{r}_{E_2} \right\| + b \\
&\;\;\vdots \qquad\qquad \vdots \\
\Delta d_{j1} &= \left\| \vec{r}_S - \vec{r}_{E_j} \right\| + b.
\end{aligned} \tag{3.38}$$

It needs to be pointed out that the indexes denoting the distance differences have been swapped. This allows to formulate the equations in an identical manner as in the TOA case. The left side can be summarized to a vector $\vec{\rho}'$:

$$\vec{\rho}' = \begin{bmatrix} 0 & \Delta d_{21} & \cdots & \Delta d_{j1} \end{bmatrix}^{\mathrm{T}}. \tag{3.39}$$

In this case, the i-th row of the new measurement vector corresponds exactly to the receiver E_i, so the double indexing used until now can be dropped. This system of equations can be converted to a linear one. To achieve this, the i-th equation needs to be expanded:

$$\rho_i' = \sqrt{(x_S - x_{E_i})^2 + (y_S - y_{E_i})^2 + (z_S - z_{E_i})^2} + b. \tag{3.40}$$

The above modification allows to solve the system using the method described by Bancroft. By subtraction of b and subsequent squaring of both sides, the following result is achieved:

$$(\rho_i' - b)^2 = (x_S - x_{E_i})^2 + (y_S - y_{E_i})^2 + (z_S - z_{E_i})^2. \tag{3.41}$$

After expanding the quadratic terms one obtains:

$$\rho_i'^2 - 2\rho_i' b + b^2 = x_S^2 + y_S^2 + z_S^2 + x_{E_i}^2 + y_{E_i}^2 + z_{E_i}^2 - 2(x_S \cdot x_{E_i} + y_S \cdot y_{E_i} + z_S \cdot z_{E_i}).$$
$$(3.42)$$

In the next step the terms connected with the MU position and the ones belonging to the AP must be grouped:

$$\left(x_{E_i}^2 + y_{E_i}^2 + z_{E_i}^2 - \rho_i'^2\right) - 2\left(x_S \cdot x_{E_i} + y_S \cdot y_{E_i} + z_S \cdot z_{E_i} - \rho_i' \cdot b\right)$$
$$+ \left(x_S^2 + y_S^2 + z_S^2 - b^2\right) = 0.$$
$$(3.43)$$

The *Lorentz inner product* in $\mathbb{R}^4$ is defined as:

$$\left\langle \vec{a}, \vec{b} \right\rangle = a_1 b_1 + a_2 b_2 + a_3 b_3 - a_4 b_4.$$
$$(3.44)$$

Using this property, equation (3.43) can be rewritten as:

$$\tfrac{1}{2}\langle \vec{v}, \vec{v} \rangle - \langle \vec{v}, \vec{u} \rangle + \tfrac{1}{2}\langle \vec{u}, \vec{u} \rangle = 0, \quad \text{with } \vec{u} = \begin{bmatrix} x_S \\ y_S \\ z_S \\ b \end{bmatrix} \text{ and } \vec{v} = \begin{bmatrix} x_{E_i} \\ y_{E_i} \\ z_{E_i} \\ \rho_i' \end{bmatrix}. \quad (3.45)$$

Furthermore, it is useful to define the relations:

$$\mathbf{B} = \begin{bmatrix} x_{E_1} & y_{E_1} & z_{E_1} & -\rho_1' \\ x_{E_2} & y_{E_2} & z_{E_2} & -\rho_2' \\ \vdots & \vdots & \vdots & \vdots \end{bmatrix}, \quad \vec{a} = \frac{1}{2} \begin{bmatrix} \langle \vec{v}_{E_1}, \vec{v}_{E_1} \rangle \\ \langle \vec{v}_{E_2}, \vec{v}_{E_2} \rangle \\ \vdots \end{bmatrix},$$

$$\vec{e} = \begin{bmatrix} 1 & 1 & \cdots \end{bmatrix}^{\mathrm{T}}, \quad \lambda = \frac{1}{2}\langle \vec{u}, \vec{u} \rangle \qquad (3.46)$$

and so the system of equations from (3.38) can be written as:

$$\vec{a} - \mathbf{B} \cdot \vec{u} + \lambda \cdot \vec{e} = \vec{0}$$
$$\Leftrightarrow \quad \mathbf{B} \cdot \vec{u} = (\vec{a} + \lambda \cdot \vec{e}).$$
$$(3.47)$$

This equation can be solved for the vector $\vec{u}$ by using the least squares method, where the $(\vec{a} + \lambda \cdot \vec{e})$ is interpreted as measurement vector:

$$\vec{u} = \underbrace{\left(\mathbf{B}^{\mathrm{T}}\mathbf{B}\right)^{-1}\mathbf{B}^{\mathrm{T}}}_{\mathbf{B}^+} \cdot (\vec{a} + \lambda \cdot \vec{e}).$$
$$(3.48)$$

46

The λ includes the unknown MU position. In the next step, the solution from (3.48) needs to be substituted into the definition of λ from (3.46). This results in:

$$\lambda = \frac{1}{2}\langle \vec{u}, \vec{u} \rangle = \frac{1}{2}\langle \mathbf{B}^+(\vec{a}+\lambda\vec{e}), \mathbf{B}^+(\vec{a}+\lambda\vec{e}) \rangle \tag{3.49}$$

$$= \frac{1}{2}\langle \mathbf{B}^+\vec{a}, \mathbf{B}^+\vec{a} \rangle + \lambda\langle \mathbf{B}^+\vec{a}, \mathbf{B}^+\vec{e} \rangle + \frac{1}{2}\lambda^2\langle \mathbf{B}^+\vec{e}, \mathbf{B}^+\vec{e} \rangle.$$

This equation can be reformulated into a quadratic equation with respect to the unknown parameter λ:

$$\lambda^2 \underbrace{\langle \mathbf{B}^+\vec{e}, \mathbf{B}^+\vec{e} \rangle}_{f} + \lambda \cdot \underbrace{2(\langle \mathbf{B}^+\vec{a}, \mathbf{B}^+\vec{e} \rangle - 1)}_{g} + \underbrace{\langle \mathbf{B}^+\vec{a}, \mathbf{B}^+\vec{a} \rangle}_{h} = 0. \tag{3.50}$$

The equation (3.50) can be solved using the p-q-formula:

$$\lambda_{1/2} = -\frac{g}{2f} \pm \sqrt{\frac{g^2}{4f^2} - \frac{h}{f}}. \tag{3.51}$$

This results in two solutions for λ where only one of them belongs to the correct MU position. The solution for the MU position $\vec{r}_S$ is obtained, when λ is inserted into (3.48). The correct solution has to be identified with the use of *a priori* knowledge. The previous position calculations can be helpful in separating the correct solution from the false one. The use of auxiliary conditions, described in section 3.4.4, can also be useful. If only one of the two obtained solutions is within the defined borders, the choice is clear. It can however happen that either both or none of the solutions fit in those borders. In this case, the residue norm would have to be used. The position solution delivering the smaller residue norm is more likely to be the correct one. Furthermore, it can happen that equation (3.51) cannot be solved for a real value of λ. In this case, the minimum of the quadratic function (3.50) would have to be calculated and used as a solution.

The modified Bancroft algorithm offers a good way of finding the starting point for iterative algorithms with minimal computational effort. As the iterative algorithms require more iteration loops in case of starting points placed far from the solution, the computation time consumed by the Bancroft algorithm will be equalized in most cases. On the one hand, as will be shown in section 3.4.7, the mBA cannot reach the level of accuracy of iterative algorithms, on the other hand, however, it requires very short computation time.

It could therefore be used stand-alone when the computational resources are limited. Another algorithm for calculation of the direct solution is presented in [CH94].

3.4.7. Performance comparison of the localization algorithms

The algorithms, presented in the previous sections, are compared with each other, to identify the one best suited for the use in a practical system. Two following comparison criteria will be used:

- the mean computational time, calculated as the average of the computational times for a set of positions

- the average error of the solution, where the quality factor is the average 3D positioning error, calculated according to (3.52),

$$\text{average 3D-error} = \frac{1}{M} \sum_{k=1}^{M} \left\| \vec{r}_{S,k} - \hat{\vec{r}}_{S,k} \right\|, \tag{3.52}$$

where M is the number of realizations (different positions), $\vec{r}_{S,k}$ is the true MU position and $\hat{\vec{r}}_{S,k}$ the estimated one.

Due to the fact that some of the evaluated algorithms have additional constraints (e.g. volume in which the feasible solution should remain) and some do not, large positioning errors can occur during the evaluation of unconstrained algorithms. This would largely afflict the average error value. For this reason the median value will also be given, as it is less sensitive against outliers. The definition of the median is given in (3.53).

$$\Delta \tilde{p} = \begin{cases} Y_{(M+1)/2} & \text{, if } M \text{ odd} \\ \frac{1}{2} \left(Y_{M/2} + Y_{M/2+1} \right) & \text{, if } M \text{ even} \end{cases} \tag{3.53}$$

In order to determine the median, the values of the dataset Y (in this case, calculated position errors) are arranged according to their size (increasing). If the number of elements in the set M is odd, the median is represented by the middle entry of the set. If the number of elements is even, the median is calculated as average value of the two middle entries of the set.
Another possibility to reduce the influence of outliers would be to sort the calculated positions from small to large, then discard around 1 % of the values at the lower and the higher end of the population, and finally calculate the mean value of the remaining components.

For the evaluation, an imaginary room with the dimensions $10\,m \times 10\,m \times 2.5\,m$ with five APs has been used. Four of the receivers have been placed in $1\,m$ height in the corners; the fifth one has been placed in the center point of the ceiling. This constellation is depicted in Fig. 3.4. The reason why this AP distribution has been selected is explained in section 3.8.

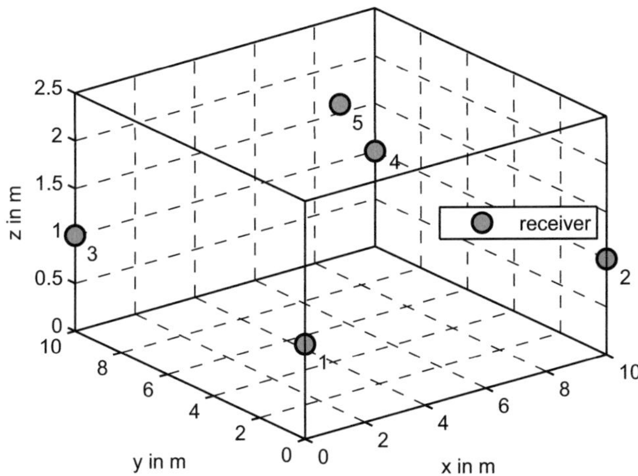

Fig. 3.4.: Scenario and AP-constellation used for the evaluation of localization algorithms.

The time information (TDOA data) for the evaluation was obtained from (3.11). The measurement noise was modeled as normally distributed, with a standard deviation of $\sigma_t = 333\,ps$ (corresponding to $\sigma_d = 10\,cm$). The evaluation of the algorithm performance is especially interesting in the presence of noise, as the real measurement values are always afflicted with uncertainty. The assumption about homoscedasticity (homogeneity of variance) of the time error at every AP is made only for the sake of algorithm testing and does not apply during measurements. The MU positions have been chosen within the scenario boundaries, depicted in Fig. 3.4, by a random function with uniform distribution. The starting point for the iterative algorithms was chosen by the modified Bancroft algorithm, to see to which extent the positioning solution improvements are possible. Using this method $M = 1000$ positions have been investigated. The results of the performed position calculations are presented in table 3.1 [18]. For the mBA and the Interior-Point algorithm the auxiliary conditions have been used, stating that the solution must be within the room.

Table 3.1.: Average 3D error and computation times of various algorithms.

Algorithm	average commutation time in ms	3D positioning error mean/median in m
modified Bancroft	0.580	0.406 / 0.312
Gauss-Newton	31.85	0.386 / 0.276
Levenberg-Marquardt	15.41	0.386 / 0.276
Trust-Region-Reflective	40.96	0.386 / 0.276
Interior-Point	98.10	0.311 / 0.251

The first impression is that the Gauss-Newton algorithm delivers good results, however, if the starting point is picked in a random manner, convergence problems occur (see also [MP06]). For this reason, the GN will not be considered from this point on. From the remaining algorithms, the mBA has by far the shortest computation time, but its accuracy is worse than that of the iterative algorithms. The LM and TRR are both in a similar accuracy range, however, the Levenberg-Marquardt algorithm requires less than half of the computation time. The Interior-Point algorithm delivers the most precise results, however, this advantage is achieved at the cost of the longest computation time of all evaluated algorithms.

Based on these results, the best combination is the starting point determination based on the modified Bancroft algorithm and subsequent final calculation with the Levenberg-Marquardt algorithm. In the case where the additional conditions regarding the geometry should be accounted for, the Interior-Point algorithm is a good choice.

3.5. Sources of errors during localization

In practice, perfect time difference measurements are never possible. Various error sources falsify the correct position estimation. The types of errors can be divided in external ones, caused by the environmental conditions, and in internal system errors. In the following two sections both types will be described.

3.5.1. External error sources

All errors that occur and influence the signal on its way from the transmitter to the receiver, and which do not originate from the UWB localization system itself, can be classified as external ones.

The most severe error source during localization is the signal time of flight delays, caused by NLOS situations. In the simplest case, the obstacle is made of a dielectric. Due to the propagation characteristic of the electro-magnetic wave through an optically more dense medium with a smaller propagation velocity, the signal time of flight becomes larger, resulting in an offset afflicted measurement at the receiver. It is also very likely that the direct path will be attenuated by the obstacle to the extent that no signal reception is possible. Signals traveling on reflection paths which arrive later at the receiver and which are less attenuated could then falsely be interpreted as signals from direct paths. This also results in a measurement time offset. Those offsets can attain significant values, depending on the scenario geometry. This observation was also done in [SGKJ07, CDW08, GP09].

In Fig. 3.5, the influence of the time offset on localization accuracy, added at different receivers from Fig. 3.4, is presented (3D positioning error). For a deeper insight, additionally the 2D and vertical error are evaluated. The MU has been positioned in the center of the room: $5\,\text{m} \times 5\,\text{m} \times 1.5\,\text{m}$, and the APs were afflicted with the time of flight offset corresponding to a distance of $[0\,\text{m},\ldots,3\,\text{m}]$. Additionally, white noise, with $\sigma_d = 4\,\text{cm}$, has been applied to time measurements at each receiver. The offsets are added to only one AP at the time, otherwise the error rises very rapidly.

It can be observed that, especially in the 2D case, the positioning error is linearly related to the applied bias. Due to the non-linear nature of the used algorithms, the calculated errors are often larger than the absolute value of the added offset. The vertical error, above a certain offset value, changes only for the 5th AP and saturates for all the others. Above this bias value for the outer APs, the change in the solution can mainly be observed in the 2D solution (no auxiliary conditions have been considered here). The 3D error, for offsets applied to the outer APs, has initially a slightly quadratic characteristic and saturates above $1.5\,\text{m}$; it runs linearly from this value on. The overall error for the 5th AP has a linear run in the investigated offset range. For larger offsets, the solution algorithms begin to show convergence problems.

3.5.2. System-dependent error sources

In this section, the errors associated with the system architecture and the way of handling the signal by the UWB localization system are described.

- Peak-detection: In reality, it is never possible to exactly and repetitively determine the time of signal reception. In the threshold detection

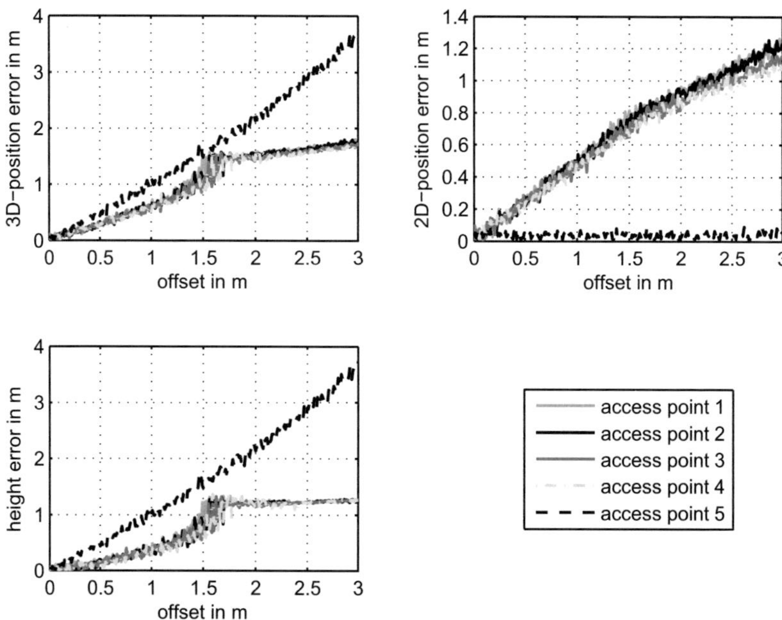

Fig. 3.5.: Investigation of the offset influence at different APs on the calculated position error. Only one access point is biased at a time.

method (cf. section 4.3.3), the time when the threshold is crossed, and thus, detection is triggered, depends on the amplitude of the received signal (see Fig. 4.15). In the case of A/D sampling, the accuracy of the arrival time estimation grows alongside with the increasing resolution and speed of the converter. In the digital domain, if the expected signal shape is known, the approximation algorithms can be used to enhance the estimation. A simulation-based example of threshold versus optimal pulse peak position detection on positioning accuracy is presented in [4].

- AP synchronization: For a precise calculation of the time differences, all the access points need to operate with the same time base. In the GPS satellite system, highly precise atomic clocks are used to ensure this, however, due to the high cost this solution is not advisable for use in the UWB system. In [HK08] it has been proposed to employ one permanently mounted transmitter for generating the timing signal for the entire localization system. However, during the transmission of the signal, similar effects to those described in the previous section can occur (e.g. NLOS), leading again to timing errors. In this situation, the simplest solution is to connect the APs by cables to create a LAN-like network. One minor drawback of this approach is that the different cable lengths, connected to the APs, need to be calibrated prior to the measurements. An architecture, very similar to this one, was used to build the demonstrator in this work (cf. chapter 6.1).

- Another error source is the thermal noise at the receiver. The thermal movement of the electrons in the receiver input stage induces a voltage, hence a constantly present noise level. The thermal noise can be well approximated by a white Gaussian noise. It is possible to determine the lower limit of the time of flight variance, in an AWGN[6] channel. According to [MOK09]:

$$(\sigma_{\hat{\tau}})^2 \geq \frac{1}{SNR \cdot \beta^2}, \tag{3.54}$$

with β denoting the effective bandwidth. For Gaussian pulses $\beta \approx 1/2\sigma_p^2$. The lower the SNR, the higher the uncertainty of the time measurement (cf. section 4.3.3). A more detailed consideration of this aspect is presented in section 6.2.2.

The last two elements of the localization system that can potentially introduce errors, are the antennas and the time measurement unit. The behavior of these devices will be analyzed separately, based on simulations and measurements, in chapter 6.

3.6. Receiver autonomous integrity monitoring

As already mentioned in the previous section, the offsets caused by the time of flight delays of the signals can cause major errors during position estimation. However, if during a measurement more APs than the required minimum are available, this redundancy can be utilized for error monitoring. For this purpose, different RAIM-algorithms (receiver autonomous integrity monitoring) can be used. The three alternatives, including the algorithm selected for this work, are described in more detail in the following sections.

---

[6]AWGN - additive white Gaussian noise

3.6.1. Least squares residual method

The least squares residual method (LSRM) can especially be utilized if a large number of APs is available. Compared to range comparison method (RCM), it requires less computational time. The idea is relatively simple: the position calculation is performed with measurements from all APs, using the algorithms described in section 3.4. In the next step, the residues of all APs are calculated according to (3.8). The identification of the erroneous AP (one or more) can be done through the assessment of the residue vector $\vec{f}(\vec{r}_S)$. In the case when a large number of receivers has been used to calculate the position, the result should not be degraded by the influence of a few APs with wrong time information. This results in small residues for error-free receivers and large entries in the residue vector in case of erroneous APs. This approach however becomes unstable for a large number of offset-afflicted APs, or a small number of APs. The later is the reason for the choice not to implement this method in this work.

3.6.2. Parity space method

The idea of the parity space method (PSM) is to use the parity space matrix for the calculation of the new error vector, whose expected value depends solely on the offsets of the time measurements. The parity space matrix can be obtained from the measurement matrix $\mathbf{H}$. The entries of the new error vector increase exactly at those positions that correspond to the biased receiver. The drawback of this approach - just as in the case of LSRM - is, that for a calculation of the new error vector the calculation of the residue vector is required. In turn, this is obtained from the position calculation using all APs.

3.6.3. Range comparison method

There is a significant difference between the approach used in the RCM and previously mentioned RAIM-algorithms (LSRM and PSM). In RCM the initial MU position calculation using all available APs is not required. Instead, multiple position estimations are performed, using subsets of only four APs, which is the minimal number of receivers required for 3D localization. Due to this advantage in stability the RCM has been implemented and is described in greater detail below. Its performance evaluation based on simulated data is presented in [3]; the one based on measured data in section 7.2.3.

Error detection

If the total number of used receivers is $N > 4$, there are in general $\binom{N}{4}$ possible position calculations. The subsets are then modified by subsequent including the initially unused APs and by calculating new position estimations. In the following step, the residuals between the positioning solutions performed with the original subset and the modified one are calculated. The position calculation has been performed here using the modified representation of the TDOA equations from (3.38) and (3.39). In this form it is possible to directly relate the AP to the entry of the measurement vector. After calculation, two cases must be distinguished:

1. An erroneous AP has been included in the initial calculation: In this case, an invalid position estimation is to be expected. As a consequence, the residuals of the unused APs are large, regardless of whether the unused APs are valid or not. In this case no identification of the invalid AP based on residuals can be done.

2. The erroneous AP has not been included in the initial calculation: In this case, the position estimation will be accurate. The residuals of the unused valid APs will remain small, whereat the residuals to the unused invalid APs will be large.

The differentiation between those two cases requires a minimum of two residuals to the unused APs. Therefore, $N \geq 6$ APs are needed. In this case it is possible to identify $N - 5$ access points with time bias. In the second case described above, the residue of the correct AP will be small, whereas the one of the erroneous AP will be large. If this is true, then after each position calculation it must be possible to detect at least one small residue. For this purpose a classification method based on two thresholds has been implemented in this work. The value of the lower threshold has been set to $s_L = 15\,\text{cm}$. Below this value the residues f_j are classified as "SMALL". The upper threshold is set to $s_U = 30\,\text{cm}$. The residues lying above s_U are classified as "LARGE". Values which fit between those two thresholds are identified as "MEDIUM".

Decision matrix

From the above description a decision matrix is created. An example of such matrix for $N = 6$ receiver is given in table 3.2. The numbers in the first column correspond to the APs used to create the primary subset. In the second column the initially unused APs are listed. Following two columns give the residue value to the unused AP.

Table 3.2.: Complete example of a decision matrix for $N = 6$ access points with corrupted receiver nr. 5.

used APs	unused APs	residue f_1	residue f_2
1234	56	LARGE	SMALL
1235	46	LARGE	MEDIUM
1236	45	SMALL	LARGE
1245	36	MEDIUM	MEDIUM
1246	35	SMALL	MEDIUM
1256	34	LARGE	LARGE
1345	26	LARGE	LARGE
1346	25	SMALL	LARGE
1356	24	LARGE	LARGE
1456	23	LARGE	MEDIUM
2345	16	LARGE	LARGE
2346	15	SMALL	LARGE
2356	14	LARGE	MEDIUM
2456	13	LARGE	LARGE
3456	12	LARGE	LARGE

The background of the decision matrix rows is color-coded. The rows with a LARGE/SMALL residue combination are marked with a dark grey color. Based on this, the faulty AP can be identified: the one that appears in all those rows (second column). In this example it is the AP 5.

Evaluation of the decision matrix
In order to evaluate the decision matrix, a counter is assigned to every AP combination that has been identified. In the case of e.g. 7 APs, the counter is assigned to every single AP, and another one to pairwise combinations of two APs. In the following step, the search for the rows of the decision matrix with at least one LARGE and one SMALL entry is carried out. If such a combination was found, then the counters for the AP combination that in this residue combination have the LARGE entries, would be increased. If 7 APs were used, and the first row looks like in table below, the counter for the combination (5,7) would be increased.

unused APs	residue f_1	residue f_2	residue f_3
5 6 7	LARGE	SMALL	LARGE

However, it would not be correct to directly identify the AP, that have the highest counter value as faulty. This is because situations where in the decision matrix only one AP residue was classified as LARGE occur much more often than a combination of two LARGE-entries. For this reason, the counter value is normalized to the maximal possible number of occurrences O_{max} of the combination R. In general, this value is given by:

$$O_{max}(R) = \binom{N-R}{4}.$$ (3.55)

Continuing the example with 7 APs results in the following: for the combination of one AP, the value $O_{max}(1) = 15$ is found and for the combination of two $O_{max}(2) = 5$. This means that exactly one AP appears in 15 rows of the decision matrix, whereat the same pair of APs (double combination) appears only in 5 rows. The normalization of the counter value c of the combination i corresponds to the calculation of the relative frequency of occurrence:

$$O_{rel}(R) = \frac{c(i)}{O_{max}(R)}.$$ (3.56)

For the relative occurrence frequency a threshold $s_{O_{rel}} = 0.9$ has been introduced. The AP combination with the highest relative occurrence frequency, whose value is larger than $s_{O_{rel}}$, is marked as faulty. Some numerical examples, of how this algorithm can improve the solution in the case of a measurement bias are given in chapter 7.

3.7. Velocity-based integrity monitoring

Another mechanism that can be used to monitor the localization solution is the observation of the velocities which result from the discrete differentiation of the calculated UWB positions [35]. Depending on the application and the scenario where the UWB localization system is used, the maximum allowed velocities for all moving objects are generally well defined. If a mobile user requires too high velocity, in order to move from the last known position to the present one, then it is very likely that the new calculated position is

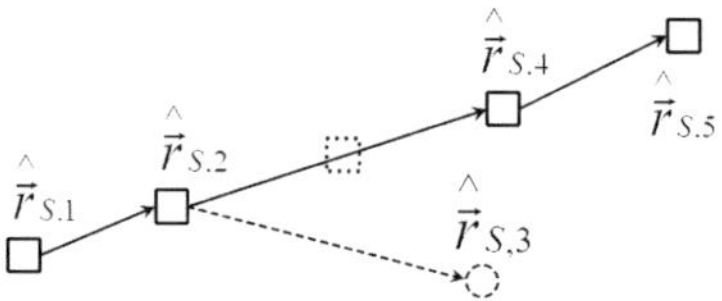

Fig. 3.6.: Detection of outliers with a velocity filter.

incorrect. The velocity can be derived from the calculated UWB solution $\hat{\vec{r}}_S$ in the following way [3]:

$$\left|\vec{v}_k\right| = \frac{\left\|\hat{\vec{r}}_{S,k} - \hat{\vec{r}}_{S,k-1}\right\|}{t_k - t_{k-1}}, \tag{3.57}$$

where $k-1$ is the point in time of the last valid position. In the case when a position is found to be invalid and is discarded, it is obvious that it cannot be used for further velocity calculations. Instead, the last valid position $\hat{\vec{r}}_{S,k-1}$ at time t_{k-1} is used, as illustrated in Fig. 3.6. The estimated position $\hat{\vec{r}}_{S,3}$ is incorrect. The threshold for the subsequent velocity filtering is adapted to include the doubled possible distance change. In this example, the velocity $\left|\vec{v}_4\right|$ has to be calculated based on positions $\hat{\vec{r}}_{S,2}$ and $\hat{\vec{r}}_{S,4}$.

The maximal velocity allowed for an MU in a certain scenario is in most cases significantly different in the horizontal and vertical direction. Therefore, two separate thresholds should be used to evaluate the plausibility of the calculated speed in those directions. In indoor scenarios (such as offices or industrial buildings) there are very few objects that move faster than $5\,\mathrm{m/s}$ in the xy-plane. The default threshold for the vertical velocity was amply set to $1\,\mathrm{m/s}$. Figure 3.7 shows the run of the estimated velocity, during a walk of a person carrying an UWB unit through an office building [Sch10]. There are two time intervals ($7\,\mathrm{s}$ to $12\,\mathrm{s}$ and $44\,\mathrm{s}$ to $51\,\mathrm{s}$) when the connection between the APs and the person is obstructed, leading to time biases. These situations can be clearly distinguished from the plausible solutions and can be sorted out by the velocity filter.

3.8. Spatial distribution of the UWB access points

Inevitably, while designing a real localization system, sooner or later the question about the optimal AP placement in the scenario has to be answered. In order to solve this issue, the question will be rephrased to: *"How does*

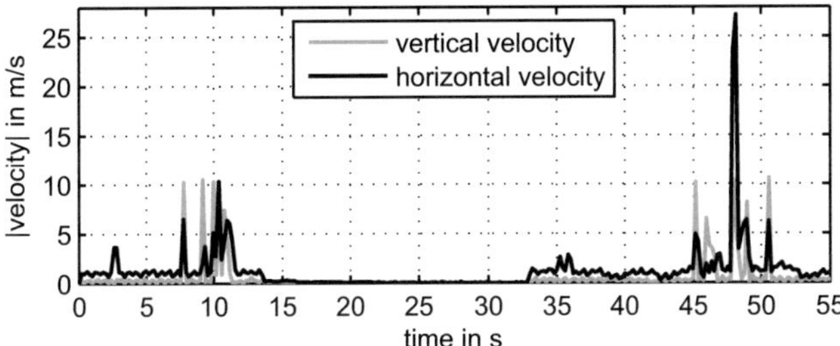

Fig. 3.7.: Change of the estimated velocity over time. Gray line: horizontal velocity, black line: vertical velocity.

an error in a time difference measurement map to the error of positioning solution?". In the following sections, this aspect of the localization system deployment will be considered from both theoretical and practical point of view.

3.8.1. Theoretical consideration based on DOP values

The DOP values are parameters that can be directly used for the assessment of the access point constellation quality. They contain the information about how strong will be the influence of the TDOA measurement noise on the position estimation. The below derivation is a result of a joint work with co-authors of [4] and published in [18].

In order to determine the DOP values, the covariance matrix of the positioning solution is calculated first. The covariance matrix of the measurement vector $\vec{\rho}$ from (3.6) is defined by:

$$\mathbf{R}_\rho = \mathbf{H} \cdot \mathbf{R}_r \cdot \mathbf{H}^\mathrm{T}, \tag{3.58}$$

and is related to the covariance matrix $\mathbf{R}_r$ of the positioning solution $\vec{r}_S$, where $\mathbf{H}$ is the linearized mapping matrix from (3.26). When equation (3.58) is solved for $\mathbf{R}_r$:

$$\mathbf{R}_r = \left(\mathbf{H}^\mathrm{T}\mathbf{H}\right)^{-1} \mathbf{H}^\mathrm{T} \cdot \mathbf{R}_\rho \cdot \mathbf{H}\left(\mathbf{H}^\mathrm{T}\mathbf{H}\right)^{-1}, \tag{3.59}$$

the general expression for the mapping of the measurement uncertainty on the localization uncertainty is obtained [LS09]. If the measurements of the signal propagation time have the same variance σ_d^2, the covariance matrix of the measurement has the following form:

$$\mathbf{R}_\rho = \sigma_d^2 \cdot \mathbf{I}. \tag{3.60}$$

In this case, (3.59) simplifies to:

$$\mathbf{R}_r = \left(\mathbf{H}^{\mathrm{T}}\mathbf{H}\right)^{-1} \cdot \sigma_d^2. \tag{3.61}$$

This is, however, only valid for uncorrelated measurements and does not apply to the TDOA case in general, as already stated in section 3.3. Although after the decorrelation the assumption about the covariance matrix of the measurement values being a diagonal matrix is valid, the time measurements have different variances. To account for this in (3.61), instead of using matrix $\mathbf{H}$ from (3.26), the decorrelated matrix $\mathbf{H}'$ from (3.29) and the weighting matrix $\mathbf{D}$ from (3.22) will be used. This results in a WDOP (weighted dilution of precision) [SAT03, BH99]. The valid formula for calculating the localization solution covariance matrix in the TDOA case has the following form:

$$\mathbf{R}_r = \underbrace{\left(\mathbf{H}'^{T}\mathbf{D}^{-1}\mathbf{H}'\right)^{-1}}_{\mathbf{Q}} \cdot \sigma_d^2, \tag{3.62}$$

where $\mathbf{Q}$ is, due to the performed decorrelation, a diagonal matrix. The entries of the diagonal give the DOP values in x-, y- and z-direction. Those are called XDOP, YDOP and VDOP (vertical dop). It is also possible to calculate the DOP values for 2D and 3D positions:

$$HDOP = \sqrt{\sigma_x^2 + \sigma_y^2} \quad = \sqrt{\mathbf{Q}_{11} + \mathbf{Q}_{22}}, \tag{3.63}$$

$$PDOP = \sqrt{\sigma_x^2 + \sigma_y^2 + \sigma_z^2} = \sqrt{\mathbf{Q}_{11} + \mathbf{Q}_{22} + \mathbf{Q}_{33}}. \tag{3.64}$$

For the calculation of the DOP values only the measurement matrix $\mathbf{H}$ and the constant weighting matrix $\mathbf{D}$ are required. As a consequence the DOPs depend solely on the geometrical configuration of the APs. Therefore, positional (3D) DOP (so-called PDOP) can be used as a quality measure of a conceptualized AP-MU constellation. In general, high DOP values indicate a poor, and small DOPs a good, AP-MU configurations.

3.8.2. Optimal access point placement in the literature and examples

The method of DOP analysis in context of the most efficient access point placement in localization systems is widely discussed in the literature. The range of investigated scenarios spans from very small ones, such as a cube with $2\,\mathrm{m}$ edge length where the AP are placed in each corner [YL10], to as large ones as a $200\,\mathrm{m} \times 140\,\mathrm{m}$ sized race track [SYG09]. The authors in [Lev00] derived the formula for the smallest achievable PDOP in a 2D scenario and state that the optimal arrangement is achieved when N APs are located at the vertices of a regular N-sided polygon. In this case the smallest DOP value is achieved in the center of the polygon. However, as concluded in [SGKJ06], the "the optimal solution (AP configuration) is only valid for one single position and not for a complete area". Because of this, it is more advisable to optimize the AP constellation to obtain a possibly low average DOP across the scenario, or at least a uniform distribution of these values. In following some example AP configurations that could be used in a typical indoor localization scenario are analyzed.

In Fig. 3.8 the distribution of the horizontal DOP (HDOP) and VDOP values for four different AP constellations is shown. The room with $10\,\mathrm{m} \times 10\,\mathrm{m} \times 2.5\,\mathrm{m}$ dimensions from Fig. 3.4 will be used here as a starting point for the consideration. The evaluation of the DOP values is performed at a constant height of $1.5\,\mathrm{m}$ and is continued up to $5\,\mathrm{m}$ beyond the borders of the room (horizontal direction), in order to visualize the significant changes of the values outside of the constellation.

The configuration C1 in Fig. 3.8(a) shows 4 receivers placed in $2.5\,\mathrm{m}$ height. The HDOP values inside the constellation are in the range $\approx [1.0, \ldots, 1.3]$ and increase up to ≈ 9 outside of it. The VDOP distribution, however, is not uniform and reaches very high values in the middle of the room (note the scaling of the color bar). At this position a vertical movement of the target would cause almost no change in the measured signal travel time differences.

This situation can be avoided by removing the symmetry of the AP configuration in the middle of the room, e.g. by adding another access point. This situation is depicted in Fig. 3.8(b). In C2, the centrally placed receiver barely influences the HDOP. At the same time reduces the VDOP range to $\approx [1.3, \ldots, 10.0]$.

The VDOPs can be further improved by shifting up the middle AP out of the plane of all other access points, or by lowering the others down, e.g. by $1.5\,\mathrm{m}$. This situation, named C3, is depicted in Fig. 3.8(c). Thanks to this

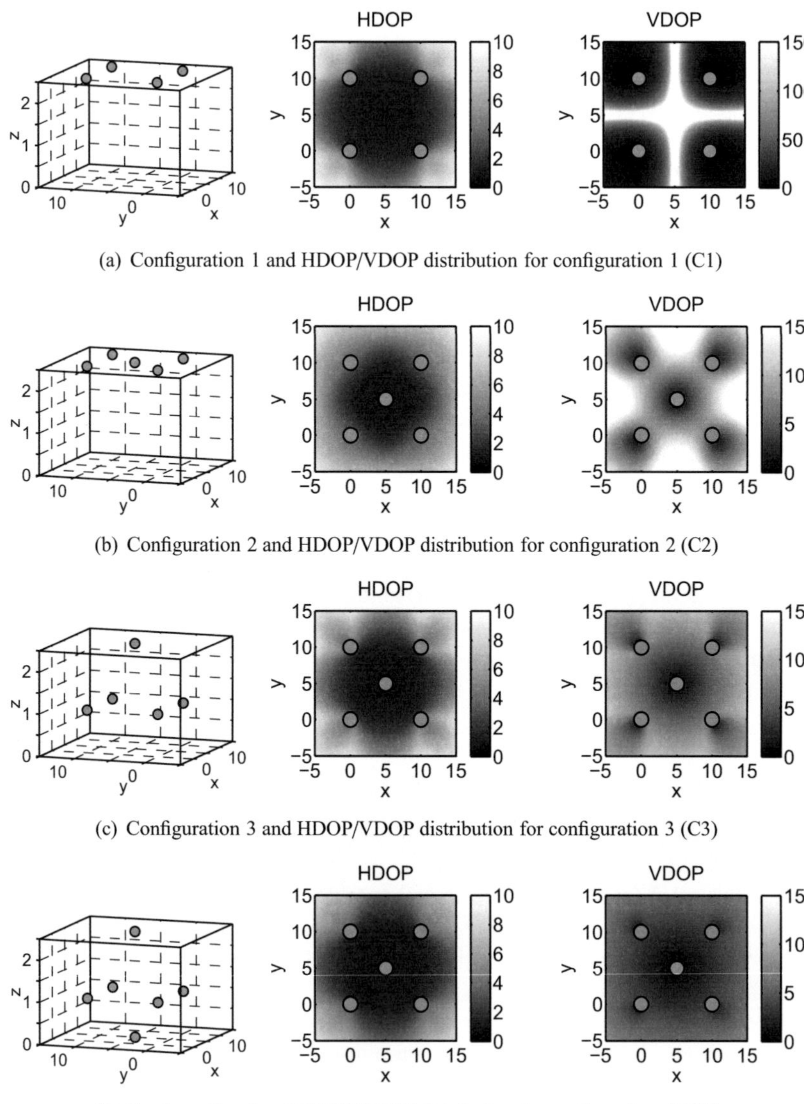

(a) Configuration 1 and HDOP/VDOP distribution for configuration 1 (C1)

(b) Configuration 2 and HDOP/VDOP distribution for configuration 2 (C2)

(c) Configuration 3 and HDOP/VDOP distribution for configuration 3 (C3)

(d) Configuration 4 and HDOP/VDOP distribution for configuration 4 (C4)

Fig. 3.8.: Various configurations of access points and corresponding DOP values. All coordinates are expressed in meters.

measure, the vertical change in position can be better resolved with VDOP $\approx [1.0,\ldots,7.0]$, whereat HDOP did not undergo any significant change compared to C1 and C2.

Another method for reducing the uncertainty in the vertical direction is to involve an additional AP placed underneath all present access points. This causes larger measured time differences between two neighboring positions (Fig. 3.8(d)), hence the VDOP $\approx [0.7,\ldots,3.7]$. It is worth noting that in all cases the horizontal accuracy of the system decreases rapidly, as soon as the transmitter is outside the receiver constellation. Therefore, the AP should be placed in a way that enables the "encasement" of the scenario in which the localization is to be performed.

3.8.3. Relation between DOP values and Cramer-Rao bound

In order to maximize the RMS accuracy of an estimator by using an optimal sensor configuration the cost function describing the estimator's performance has to be minimized [Nee09]. In the previous section one possible method has been described, namely the minimization of the diagonal elements of the PDOP matrix. The other possibility is the minimization of the Cramer-Rao lower bound (CRLB; described in section 6.5), or equivalently maximization of the diagonal elements of the Fischer Information Matrix. The application of this approach for finding the optimal AP configuration was demonstrated e.g. in [JDW08] for TOA ranging system or in [LS09] for 2D TDOA localization.

In turn, those methods can be used for approximation of the localization error at a given time measurement noise and AP constellation. If the employed estimator is efficient [PAK$^+$05, Nee09] and the independent and identically-distributed Gaussian noise on TDOAs is assumed, the WDOP matrix attains the Cramer-Rao lower bound [CA94]. In general, however, the CRLB gives only the lower limit of the localization error (unbiased and efficient estimator), whereas the DOP approach can deliver localization error predictions closer to reality. This is because, according to [SGKJ06], the PDOP can be defined more generally than in (3.63), by:

$$PDOP = \frac{\text{RMS position error}}{\text{RMS TDOA error}}, \qquad (3.65)$$

where the RMS TDOA error does not have to be necessarily zero-mean. In absence of time bias the standard deviation of position error equals $\sigma_{\text{pos 3D}} = \text{PDOP} \cdot \sigma_{\text{TDOA}} \cdot c_0$. In [BH99] this relation has been called the "RMS spherical uncertainty".

3.8.4. **Practical aspects of the AP placement**

In theory, an almost infinite number of different configurations could be tested. In this section however, the AP placement is considered taking into account practical aspects. Both AP placements: on the ceiling (C2) and walls (C3), are feasible in practice. Despite of the non ideal VDOP, the C2 has the advantage that potential problems due to shadowing effects can be minimized, as in the majority of indoor scenarios most objects are placed on the floor or close to it.

According to the theoretical investigation performed in section 3.8.1, the installation of at least one AP at the floor level (C4) is required to achieve a significant improvement in the height accuracy. However, due to e.g. tables or mobile file cabinets in office buildings, strong signal reflections and shadowing are to be expected, which in most cases would render the information contained in the signal time of flight useless. One should however keep this possibility in mind, as scenarios in which the geometrical conditions do allow the use of this constellation may exist. One possible trade-off is the installation of one of the APs at the half of the room height. This would lead to an acceptable height accuracy, while the feasibility of the practical realization is maintained.

The question about the minimal required number of APs strongly depends on the architecture of the building to be equipped with the UWB localization system. In the presence of thick walls that cause significant attenuation and additional propagation delay of the signals, it may be required to place a minimum of four APs per room. The lightweight construction of the buildings that often can be met in the US could possibly allow to transmit UWB impulses through walls without significant amplitude loss and delay. In this case, it would not be necessary to equip every single room with four APs, required for 3D localization. The number of access points could be further reduced by merging the UWB localization with an inertial navigation system, as described later in section 7.3.

At the same time it has to be mentioned that a larger number of APs would give a rise to a more uniform distribution of the DOP values and better performance in terms of shadowing. On the other hand, the cost of a practical implementation would increase.

Based on the considerations presented in this and previous section, it is advisable in practice to arrange the APs as in the configuration C3, or a similar one. The APs, used in chapter 7 during measurement verification, have been ordered taking this conclusion into account.

4. Hardware implementation and receiver performance comparison

In the last years several UWB demonstrators for communication and localization purposes have been built [Sto08, DSD$^+$07]. However, in the majority of cases the choice of the transceiver architecture is based on the theoretical analysis, that can be found in the literature (e.g. [Ree05], ch. 6). Also more practice oriented receiver comparison criteria have been proposed, such as e.g. energy consumption [VD08] or the achieved bit-error-rate [DB05]. The most universal criterion, however, would be the evaluation of the signal-to-noise ratio in the baseband signal, before the A/D conversion. Though up to date, no work describing such performance verification and backed up with measurement data have been published. Therefore, in this thesis a universal comparison criterion for evaluation of the SNR performance of IR-UWB receivers in time domain has been developed and verified.

This chapter begins with an overview of UWB transmitter and receiver architectures that can be found in the literature. Their advantages and disadvantages are briefly discussed and based on this, three receiver topologies are selected for a more detailed investigation. The comparison between them is based on simulation and measurement. A fair comparison is assured by the common hardware components developed in this thesis, that are used to build the selected receivers. The evaluation criterion is the threshold-to-noise ratio (introduced in chapter 2), that is investigated at multiple transmitter-receiver distances.

4.1. Impulse-radio transmitter topologies

Impulse-based UWB systems are characterized by a very simple transmitter stage, that in most cases generates carrierless signals. In this section, the most common impulse radio UWB transmitter types are revised and briefly characterized. Further, an universal transmitter architecture is introduced.

4.1.1. Most common Tx structures

One of the most simple, and hence, popular method of generating UWB impulses uses a two-stage architecture. Firstly, a circuit that will convert the trigger signal into rapidly rising voltage or current signal is employed. Several techniques have been used to achieve this: avalanche transistors [RZ75], biased diodes (e.g. step-recovery-diode [RB04]) or logic gates [KPJ04]. In the second stage, depending on the regulatory issues, the pulse needs to be shaped; this usually involves shorted- [Sit09] or opened-stub [PTZW09] filter-like networks. This approach is schematically depicted in Fig. 4.1(a).

Another, more advanced and compact, solution is depicted in Fig. 4.1(b). Here, the pulse generation is done with one integrated device, without any further need for shaping. It is possible to employ e.g. a limiting amplifier to obtain a signal with very steep flank and use a subsequent LC-network to form the pulse [DTS06b]. The pulse is already in the transmission band and can be directly radiated through the antenna, after optional amplification.

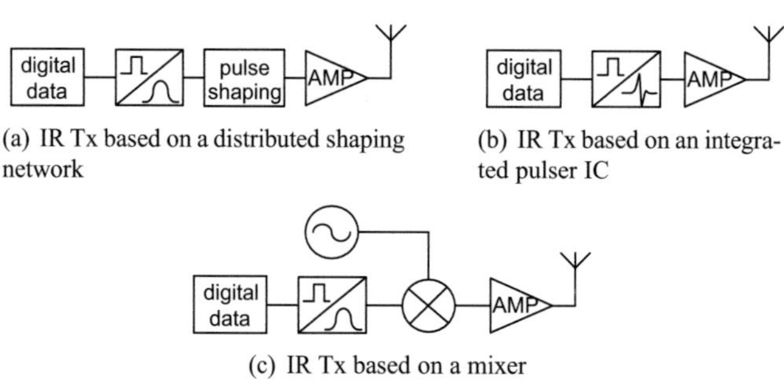

(a) IR Tx based on a distributed shaping network

(b) IR Tx based on an integrated pulser IC

(c) IR Tx based on a mixer

Fig. 4.1.: IR-UWB transmitter architectures.

The third alternative is the solution known from narrowband systems: pulse generation in baseband followed by an up-conversion stage (cf. Fig. 4.1(c)). This approach, although simple, has the drawback of significantly higher power consumption than the other two, and in the majority of cases, the issue with LO leakage to the RF output, thus violating the spectral emission limits [DDGC11].

4.1.2. Universal UWB transmitter

There exist numerous modifications of the above architectures. This short summary is intended to give an insight in this topic. Since one of the goals of this thesis is to compare the performance of several UWB receiver types (see section 4.2.5), it is reasonable to use only one transmitter hardware to ensure a fair comparison. For this purpose, short pulses of several hundred ps in length will be generated by a dedicated pulser circuit (PG), triggered by a baseband signal. The pulse can either be amplified (Amp.) or be sent directly to the antenna. This setup is depicted in Fig. 4.2. The variable attenuator is required to synthesize different SNR levels (cf. section 4.4.2) at the output of the Tx. The part of the circuit marked with the dashed line is required for the operation of a particular receiver type and will be described later in section 4.2.3.

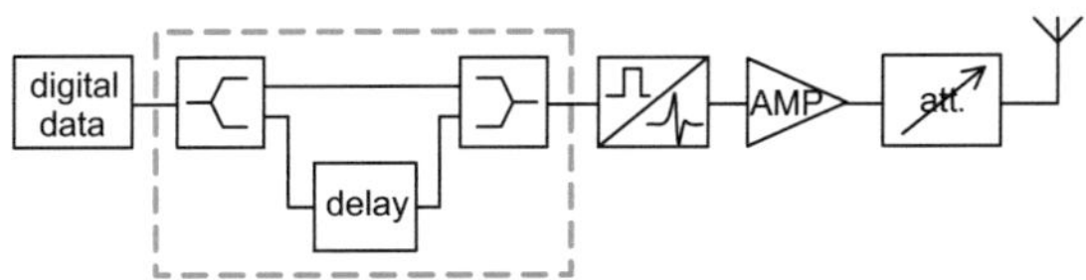

Fig. 4.2.: Universal UWB transmitter concept for correlation and auto-correlation transceiver types. The additional part, marked with the dashed box, is required for the transmitted-reference system.

4.2. Receiver Topologies

In this section, a summary of receiver architectures that can be encountered in the literature is given. Based on this summary, the candidates for closer investigation are chosen.

4.2.1. Fully and partly digital receivers

Digital receiver

A digital receiver consists only of a low noise amplifier, band-pass filter and ADC, similar to the one depicted in Fig. 4.3. The true fully digital receiver, where the ADC is placed directly behind the antenna is hard to realize in the UWB band due to very low signal power. The main difference among the receivers of this type concerns the resolution of the used ADC. With this, the signal detection is carried out in the digital domain.

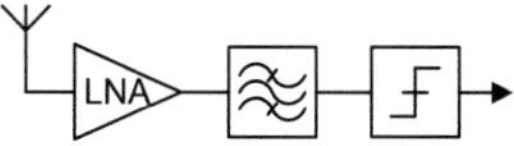

Fig. 4.3.: Digital UWB receiver with comparator as digitizing unit (1 bit ADC).

Flashing receiver (FR)

A receiver exhibiting even lower complexity than the one described above is the UWB "flashing" receiver [TII+06]. This receiver consists mainly of two comparators that continuously search for positive or negative pulse peaks in the incoming signal. This receiver type does not have a purely digital architecture, as the comparators (1 bit ADCs) are placed directly after the gain stages. The decision regarding the pulse detection is made in the digital domain, based on a recognition of a predefined pattern, related to the pulse shape.

Threshold (1-bit ADC) receiver

This architecture can be realized for example with a tunnel diode, an avalanche transistor [Ree05] or high-speed integrated comparator [ZL08]. This is the simplest type of receivers, however its usage is limited to the applications where the pulse powers are high (e.g. the military radar systems). If operated in the low SNR region, this receiver is prone to false signal detection.

4.2.2. Energy detection receivers

Rectifier receiver

A rectifier receiver can be either realized in a half- or full-wave configuration (HWR/FWR). The configuration with a pre-amplifier and bandpass filter is represented in Fig. 4.4(a) (the diode symbol does not indicate how it is connected in the circuit). Before lowpass filtering, the negative voltage parts of the signal are either cut (HWR) or turned positive (FWR). This requires the diodes to be properly biased, otherwise the sensitivity of this receiver will be poor. This can be easily done in the case of a half-wave rectifier, however the part of the energy stored in the negative voltage of the pulse will be lost [SGTS09]. In case of the HWR, pulses with the same energy content but

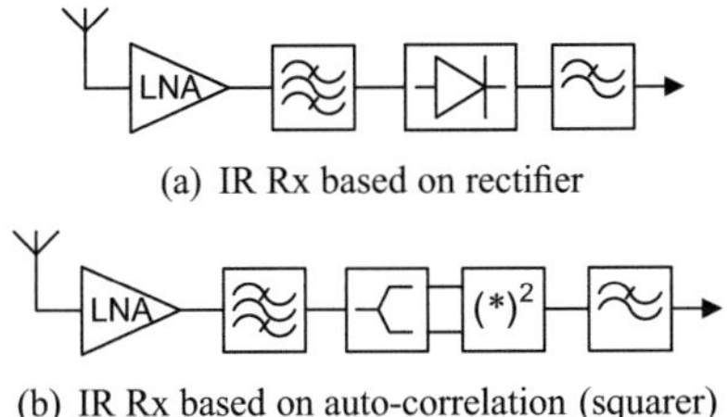

(a) IR Rx based on rectifier

(b) IR Rx based on auto-correlation (squarer)

Fig. 4.4.: Non-coherent, energy detection receiver architectures.

different shapes (e.g. positive and negative components) will induce different signal levels behind the lowpass filter.

Auto-correlation receiver (ACR)

The ACR is an active-circuitry-based alternative to the full-wave rectifier, in which no diode biasing is necessary. This concept is based on an analog multiplier. The received signal is split and squared (Fig. 4.4(b)). The main drawback of this solution is the fact that the noise is multiplied with itself, hence causing the noise floor to rise. The resulting noise signal at the output of the FWR or ACR is not zero-mean any more.

4.2.3. Coherent receivers

The term coherent, in context of UWB reception, refers to the method of pulse recovery based on correlation with a template signal. Coherent receivers can be classified based on the template signal that is employed. The so-called *clean* template is a type of signal that is locally generated in the receiver. The shape of a clean template is identical with that of the transmitted pulse. On the other hand, the *dirty* template accounts for the signal distortion in the radio channel, acting as matched filter and maximizing the correlation gain.

Correlation receiver (CR)

This receiver architecture, depicted in Fig. 4.5(a), is based on an analog and time coherent correlation of the received signal with a locally generated template. In theory, it is the most effective reception method in the AWGN channel [Tim10]. After multiplication with the clean template, the noisy received impulse generates an output, that is much larger than the correlation result of the template with received noise only (in absence of the UWB pulse). Due to

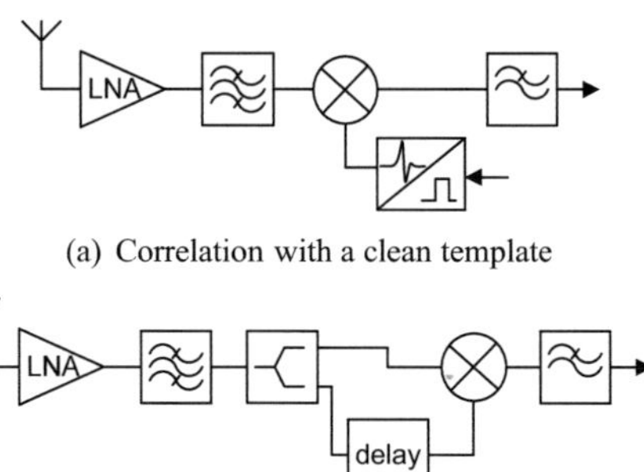

(a) Correlation with a clean template

(b) Transmitted-reference scheme

Fig. 4.5.: Coherent IR-UWB receiver architectures.

the very short duration of the UWB pulses, the requirements on the timing precision are very high (in the range of $\pm15\,\text{ps}$, depending on the pulse shape [CZ07],[15]).

Transmitted reference (TR)
In this configuration, two pulses are transmitted with a predefined relative delay τ_{TR} (usually on the order of 1 ns to 2 ns). This is achieved by triggering the PG as marked in Fig. 4.2 with a dashed line. The signal in the receiver is divided and one of the paths is delayed exactly by τ_{TR} (Fig. 4.5(b)) [HT02]. In this way, the delayed copy of the first pulse (so-called pilot signal) is correlated with the second pulse (dirty template correlation). The two other pulses resulting from the splitting (undelayed part of the first and delayed one of the second) are multiplied with noise. The main advantage of this architecture is that the time-shifted thermal noise is not correlated with itself any more, and the identical template is used for correlation. The cross-correlation of the signal and noise can however raise the noise floor as well. Another drawback of this concept is the need for realization of broadband analog delay lines in the receiver, that can attenuate and distort the pulse [TLL$^+$09]. A more detailed analysis is presented in section 4.4.1.

4.2.4. Quadrature analog correlation receiver

The quadrature analog correlation receiver (QACR) is based on the carrier recovery principle and is depicted in Fig. 4.6. The received pulse is split into two branches and down-converted to baseband with in-phase (I) and quadrature (Q) components that are then A/D converted [VD08, BZS$^+$05]. This scheme is not UWB-specific as the pulse detection is performed after the ADC (evaluation of I and Q components in digital domain). This approach does not allow for the direct comparison with other structures described before. Besides, this architecture puts tremendous requirements on the ADC. In the recent years some fast ADCs with low energy consumption, have been realized for the lower UWB band (up to 960 MHz) [GBM10]. However, at the moment, this is still unlikely to arrange this with a third contradictory demand: the one for high resolution [LHB10].

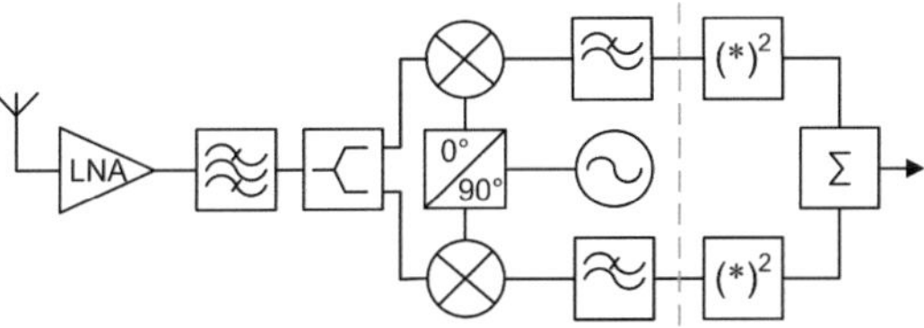

Fig. 4.6.: QAC / IQ receiver architecture. The gray dashed line represents the optional analog-digital interface.

4.2.5. Candidates for comparison

As previously described, due to the non-ideal hardware it is barely possible to assess which system has the best performance purely by mathematical consideration. Based on the overview and the considerations conducted in the previous sections, the CR, ACR and TR architectures are chosen for further detailed investigation. The reason for this choice is their simplicity and the inherent broadbandness of these topologies. In addition, they have many common components, such as: the LNA, power splitter, correlator (true multiplier), low- and bandpass filter, allowing a fair comparison by performing only simple reconfiguration. For the investigation, all three transceivers have been built from the same components [10] and their performance has been evaluated [14].

4.3. Hardware implementation of
common transceiver components

In this section, the common RF hardware components needed for the construction of the transmitter unit and all three receiver types are introduced.

Although the baseband elements introduced in section 4.3.3 will not be directly involved in the performance comparison presented at the end of this chapter, they will however be indispensable for the operation of the demonstrator presented in chapters 5 and 7.

4.3.1. Transmitter RF-frontend components

UWB impulse generator

The task of the impulse generator is to convert the digital baseband signals (square wave, bipolar) into broadband RF pulses. This form of the signal is used to spread the energy of the signal over a wide bandwidth making it possible to fulfill the signal PSD requirements according to UWB regulation. The pulses, after the PG can either be directly radiated by the antenna or their amplitude can first be adapted (amplified or attenuated). In the case of the correlation receiver, the same circuit is required at the receiver site to generate the clean template.

In this project, an integrated PG IC, presented in [DTS06b], is used to generate the RF signal, which shape is close to the one calculated in section 2.1. Bond wires have been used to connect the IC with the PCB[7].

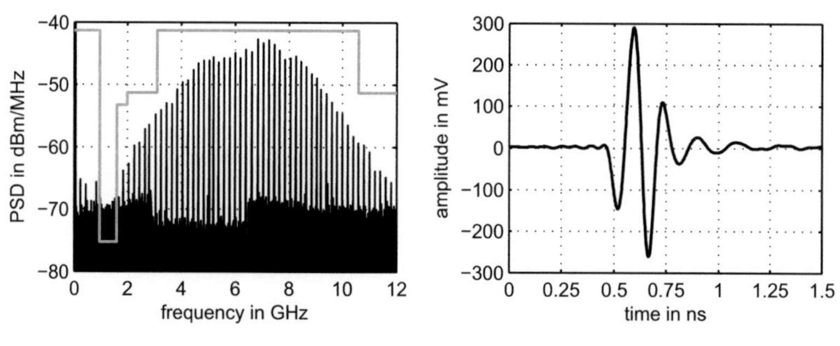

(a) Measured PSD at 200 MHz trigger signal   (b) Pulse shape in time domain

Fig. 4.7.: Frequency- and time-domain characteristic of the impulse generator circuit.

The PG operates at 3 V and consumes a total power of 38 mW, while transmitting the pulses at 200 MHz repetition rate. Such a low power consumption allows for battery operation. Due to the typically very small duty cycle of UWB pulses, it can be further reduced by introducing a sleep mode [SDS09].

---

[7]PCB - printed circuit board

The output of the PG is single-ended and matched to 50 Ω. The single-ended input needs to be DC-decoupled and is best suited for operation with digital signals (square wave) with sharp flanks, with a maximal peak-to-peak voltage of 0.4 V. At each falling edge of the trigger signal one impulse is generated. The pulse has a width of ≈ 270 ps FWHM and a -10 dB bandwidth of ≈ 7 GHz. The maximal amplitude of the output impulse is 500 mV, which can be slightly adjusted through the tuning input (DC voltage between 4 V and 6 V). The output impulse of the PG, measured in the time domain, and the PSD of the pulse train at 200 MHz PRF, are depicted in Fig. 4.7(b) and 4.7(a) respectively.

Tx amplifier

In order to bridge distances of more than only a few meters and achieve better SNR at the receiver, the transmitted impulse can additionally be amplified prior to radiation over the antenna. The main requirements on the amplifier are a flat gain in the frequency band of interest (3.1 GHz to 10.6 GHz) and a linear characteristic of the group delay; this ensures the least possible distortion of the transmitted impulse.

In this project a HMC753LP4E low noise amplifier from Hittite Microwave Corporation has been used to amplify the impulses. This module offers a gain of 17 dB and a noise figure of ≈ 2 dB. The 1 dB-compression point of 18 dBm allows to use this LNA with the previously described PG, without operating in the saturation region.

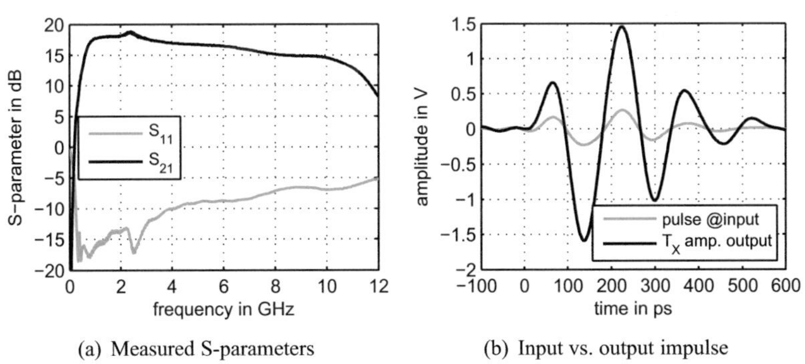

(a) Measured S-parameters   (b) Input vs. output impulse

Fig. 4.8.: Frequency- and time-domain characteristic of the Tx-amplifier.

The measured scattering (S) parameters of the amplifier are presented in Fig. 4.8(a). The non-optimal input matching, lying partly above -10 dB, causes the reflection of the impulse. By placing the PG and the amplifier closely one after another (photo see Fig. 4.16(a)) the reflected pulse travels back and forth between those two elements, causing a ringing effect. The amplified pulse in the time domain is depicted in Fig. 4.8(b). Ringing can cause problems during the correlation reception, as the pulse gets broader (more sidelobes) than the clean template. In consequence, the synchronization can be less exact (cf. section 5.1). In contrast, the quality of reception in the case of ACR and TR types of receivers should not degrade, as the pulse is multiplied by itself. Fig. 4.16(b) shows the power supply of the hybrid PG-amplifier module.

Variable attenuator

The variable attenuator can be used to adjust the signal amplitude (cf. scheme from Fig. 4.2). This can be necessary in two cases:

- The adjustments of the pulse amplitude give the possibility to control the average and instantaneous power. This allows to stay within the peak and average power limits and hence optimally using the resources, for example while changing the PRF.

- In a static scenario, with transmitter and receiver at a constant distance and operating with constant PRF, change of the Tx amplitude will cause a change in the SNR level at the receiver. This can be used to investigate the behavior of different receiver types in various SNR regimes, without the need to alter the Tx-Rx distance. This will be explained in more detail in section 4.4.

The attenuator used for this purpose is the HMC425LP3 6 bit digital attenuator. The attenuator bit values are 0.5 (LSB), 1, 2, 4, 8, and 16 dB for a maximal total attenuation of 31.5 dB. This module requires 5 V supply voltage. The measured S-parameters are shown in Fig. 4.9(a). The insertion loss is at the level of around 3.6 dB (at 6 GHz) with a ±1 dB flatness. The return loss remains below −10 dB within the bandwidth of interest. The influence of the attenuator on the pulse in time domain (Tx amplifier output) is shown in Fig. 4.9(b).

4.3.2. Receiver RF-frontend components

Low noise amplifier

The task of the LNA is to amplify the received impulse, which has been attenuated and distorted during propagation through the wireless channel. At the same time it is desired not to degrade the SNR. The LNAs used in

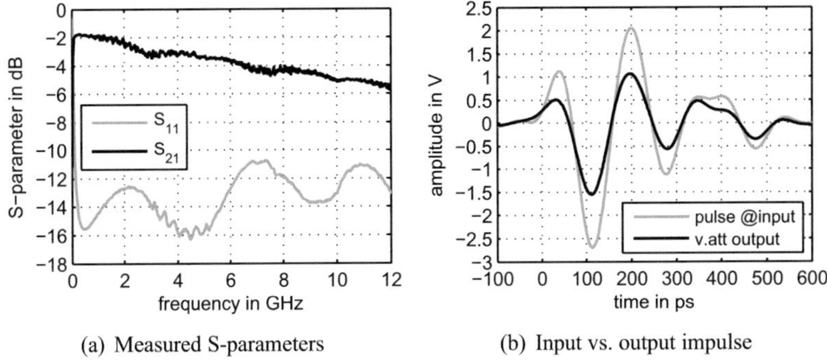

Fig. 4.9.: Frequency- and time-domain characteristic of the 6-bit digital attenuator.

broad or ultra-broadband applications need to exhibit a flat gain over the entire bandwidth and a linear phase response in order to obtain a constant group delay. This feature is especially important in the case of the correlation receiver, as it leads to the least impulse distortion.

In this project an IC presented in [DTS06a] has been taken as the basis for the LNA module. This LNA is especially sensitive to parasitic inductance caused by the bond wires. This requires special care during the connecting of the IC, including the embedding the module in a substrate cavity and using very short bond wires. Due to the relative bandwidth of around 100 %, narrow-band compensation techniques cannot be applied here. Within this thesis, dedicated LTCC[8] packages have been developed (for LNA, PG and also multiplier), which significantly simplified the assembly and assured more reproducible results [Kro12].

In Fig. 4.10(a) the measured S-parameters of the amplifier are presented. The gain of the LNA has a flat run at 22 dB with a gain flatness of approx. ±0.5 dB, starting close to DC all the way up to 10 GHz. The return loss is better than −10 dB in the entire band. The dashed curves represent the on-wafer measurement of the LNA IC and the solid ones of the LTCC packaged module, soldered on a test-PCB (photo see Fig. 4.16(d)). The noise figure of the LNA mounted on the PCB is less than 3 dB. The time-domain behavior of the LNA is depicted in Fig. 4.10(b).

Bandpass filter

The analog bandpass filter (BPF) is responsible for limiting the bandwidth of the received signal.This allows the improvement of the SNR by rejecting the

---

[8]LTCC - low temperature co-fired ceramic

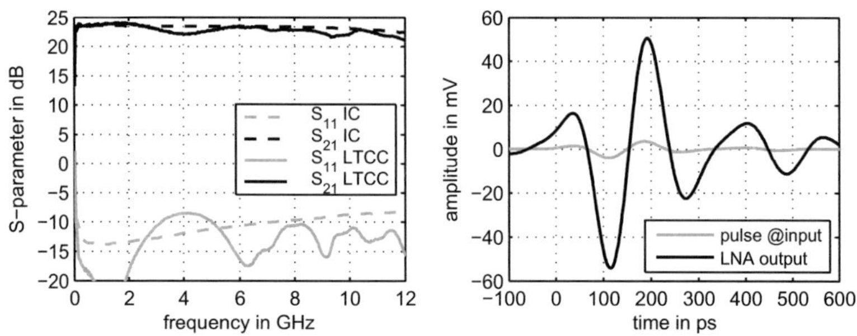

(a) S-par.: on-wafer and packaged version   (b) In- vs. output impulse (packaged version)

Fig. 4.10.: Frequency- and time-domain characteristic of the LNA.

out-of-band noise, as well as out-of-band signals coming from other wireless systems (e.g. GSM or WLAN). This kind of filter could as well be incorporated at the transmitter site to ensure that the radiated signals remain within the allowed bandwidth.

The BPF designed for the integration in the UWB receiver is quarter-wavelength short-circuited stub filter of the Chebyshev type. This type of filter has the property of being wideband without exhibiting immediate spurious bands beside the desired passband and having physical dimensions allowing integration on a PCB. In addition, Chebyshev filters have the characteristics of a steeper roll-off compared to maximally-flat filters. The 7th order Chebyshev filter satisfies the requirement for passband ripples less than 0.05 dB and minimum stopband attenuation of 40 dB. More details on the design procedure of this filter and its further development can be found in [Sit09] and [Kro12], respectively. The realized filter is depicted in Fig. 4.16(c).
In Fig. 4.11(a) the measured S-parameters of this FCC-compliant BPF, designed for the requirements of this project, are presented and compared with the EM simulation of the model done in CST [Com12]. Within the band of interest the return loss remains mostly below -15 dB and the insertion loss has a flat run at the level of only -1 dB. Towards higher frequencies, the filter attenuation increases, mainly due to the higher substrate losses.

The time-domain characteristic is presented in Fig. 4.11(b) and indicates that nearly no pulse distortion. There is also no ringing effect, what makes this filter is well suited for IR-UWB operation. The filter causes only slight broadening of few ps in the back part of the pulse.

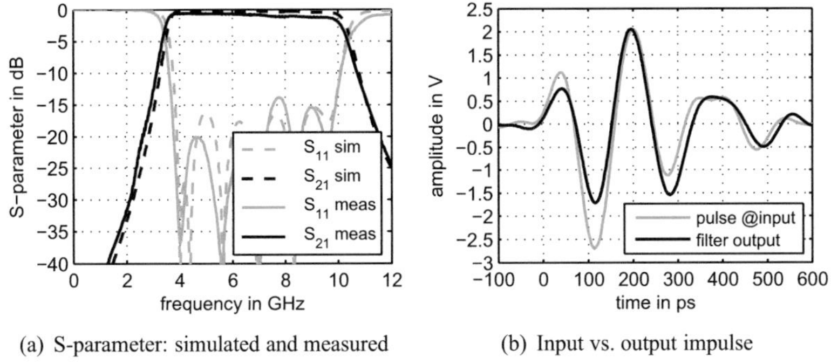

(a) S-parameter: simulated and measured   (b) Input vs. output impulse

Fig. 4.11.: Frequency- and time-domain characteristic of the bandpass filter.

Wideband power divider

Another challenging element in the receiver chain that needs to be designed is the broadband power divider. This component is required in both ACR and TR receiver structures and, due to the nature of the used UWB impulses, it needs to operate over almost two octaves of frequency. As described in [Coh68], a multistage Wilkinson power splitter can perform well over more than a decade of bandwidth. The bandwidth of this 3-port can be improved by increasing the number of sections. In [CP09] the power divider with 7 sections has been presented. Additional sections however lead to an increase of insertion losses and the size of the device. For the application presented in this thesis, a 3-stage Wilkinson divider was identified to offer the best trade-off between complexity, size, losses and bandwidth. The design is based on [LLB84]. This passive network, due to the reciprocity theorem, can as well be used as a power combiner.

In this work, for the design of the 3-stage Wilkinson power divider, the center frequency f_M has been selected, being the arithmetic mean of the upper and lower cut-off frequency, specified by the FCC. Although the calculated center frequency of the FCC UWB band is 6.85 GHz the design process was performed at 7 GHz, to account for the slightly lower effective substrate permittivity in practice, than the one theoretically calculated. The lower permittivity results in a larger wavelength, shifting the characteristic towards lower frequencies.

The impedance values of single sections and isolation resistors were obtained using the formulas from [Coh68]. A model of the power divider has been created and simulated in CST software and fabricated on a 0.5 mm thick

Rogers 4003c substrate. In Fig. 4.12 the simulation and measurement results are presented. In the measurement results at the frequency of 8.8 GHz the transmission loss is smaller than $-3\,$dB and thus better than theoretically possible. Due to the fact, that this behavior can be observed at both ports a fabrication error can be ruled out. A small calibration error is more probable. Both insertion and return losses show the expected behavior. S_{11} remains far below $-10\,$dB within the desired frequency range and the additional signal attenuation is at a level of $-0.5\,$dB. The isolation between the two output ports was measured and is better than $-15\,$dB within the relevant bandwidth.

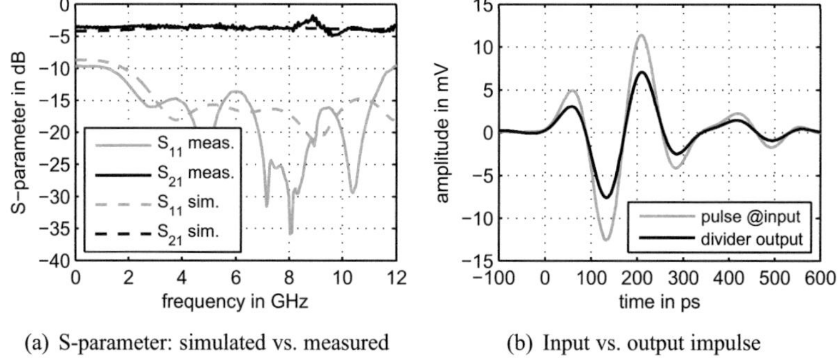

(a) S-parameter: simulated vs. measured   (b) Input vs. output impulse

Fig. 4.12.: Frequency and time domain characterization of the designed Wilkinson power divider.

In Fig. 4.12(b) the time-domain behavior is depicted. The designed divider splits the input pulse equally between two ports, without distorting its shape. The photograph of the realized power divider, incorporated in the ACR module, is shown in Fig. 4.16(e).

Broadband multiplier

The key element of the UWB receiver is the broadband multiplier which enables the implementation of the analog correlation. This operation carries out the impulse detection in the analog domain and reduces the signal bandwidth behind the correlator (transition to baseband), without reducing the amount of information carried by the signal.

The IC used to implement the analog correlator is a four-quadrant true multiplier, presented in [DST$^+$08]. The multiplier circuit has two inputs: one for the reference signal (i.e. the template impulse), and another for the received signal. The second branch incorporates in addition the LNA module, as des-

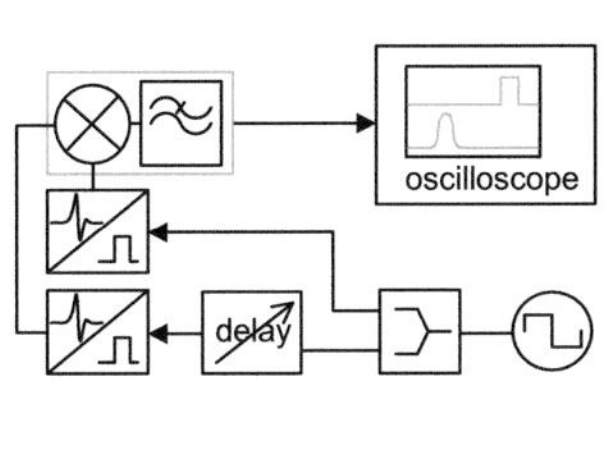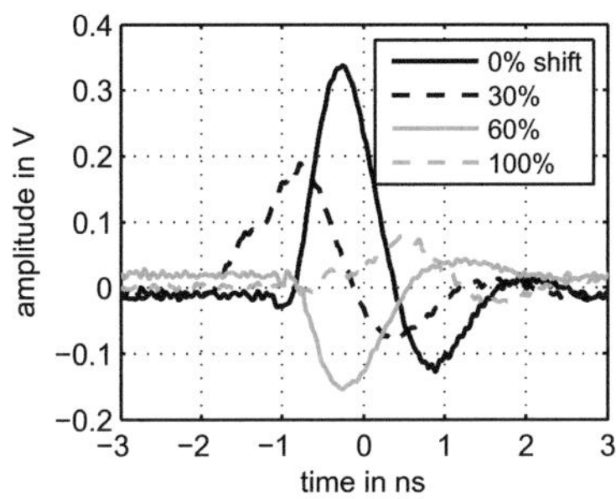

(a) Multiplier measurement setup   (b) Correlator output at four input phases

Fig. 4.13.: Measurement configuration used for the characterization of the multiplier and the output of the multiplier, depending on the template pulse position. Template shift is expressed in percent of the pulse width, relative to optimal overlap.

cribed before. Both inputs are matched to $50\,\Omega$. The output of the multiplier is differential (high ohmic). The output $3\,$dB-bandwidth is limited to $900\,$MHz by the used on-chip buffer stage, that acts as a low-pass filter (LPF). It has to pointed out that, according to the consideration presented in section 2.3.3, filtering with too low cut-off frequency reduces the STR in the output signal. A low-pass filtering at $\approx 2.5\,$GHz would yield $\approx 3\,$dB improvement in terms of STR.

In order to connect the multiplier with the digitizing unit (see section 4.3.3) and to enable measurements with conventional single-ended equipment, the correlator is connected to an operational amplifier (AD8000 from Analog Devices), that is operated in the differential mode. This particular op-amp truncates the bandwidth further to $500\,$MHz at an amplification ratio of 6.5 (linear scale). The multiplier circuit consumes $200\,$mW at $3\,$V supply voltage.

In Fig. 4.13(a) the measurement setup, used for the time-domain characterization of the multiplier, is depicted [12]. Two PGs are connected to the multiplier inputs, where the trigger signal phase of one of them is controllable by a tunable delay line. This allows the investigation of the multiplier output at different relative positions of the pulses at both inputs. In Fig. 4.13(b) the correlator output for four different relative pulse positions is shown. The maximum output amplitude changes depending on the shift and follows the curve of the auto-correlation function (ACF) (cf. section 5.1). A picture showing the multiplier and the adjacent baseband amplifier, integrated in the auto-correlation receiver module, is presented in Fig. 4.16(e).

4.3.3. Analog-to-digital conversion and digital pre-processing

In this section, the detection method of the analog baseband impulse, which comes out of the correlator is described. This process can be divided in two stages, described below: the digitization of the signal and its conditioning. The second module is required to perform receiver synchronization tests, described in chapter 5. It is also required for the localization demonstrator from chapter 7. The block diagram of both connected elements is presented in Fig. 4.14 and explained further in this section.

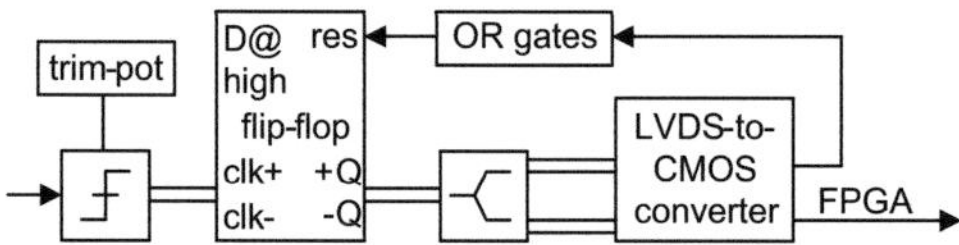

Fig. 4.14.: Block diagram of the pulse digitization unit, consisting of the 1 bit comparator and the pulse stretching circuitry.

Comparator and threshold detection

At an interface, between the analog and digital domain, both amplitude and time errors can appear. Depending on the digitizing device (comparator or ADC with more than one bit resolution) the amplitude error has different values. Obviously the ADCs with 8 bits, or even more, are capable of transforming the signal into the digital domain with only marginal amplitude distortion. The problem with high resolution ADCs nowadays is their limited bandwidth. Therefore, and due to their cost and power consumption (e.g. pipe-line ADCs consume over 1 W), they will not find application in UWB systems for the mass market. Comparators are less accurate, but by far cheaper and are a better alternative for this application. These devices can achieve input bandwidths close to 10 GHz and equivalent input signal rise times of 80 ps [Kol11].

One of the issues when using a comparator is the choice of the threshold level. In a scenario with large dynamic range the trigger time dependency on the signal level will play an important role. This is depicted in Fig. 4.15. The comparator in the UWB receiver triggers on the rising pulse edge, hence the detection time point is offset with respect to the position of the true pulse maximum. The trigger offset varies depending on the pulse amplitude. In a localization system, this situation (depicted in Fig. 4.15) would lead to erroneous TDOA readings. The optimal solution would be the use of an adaptive threshold or an automatic gain control (AGC) stage in the receiver.

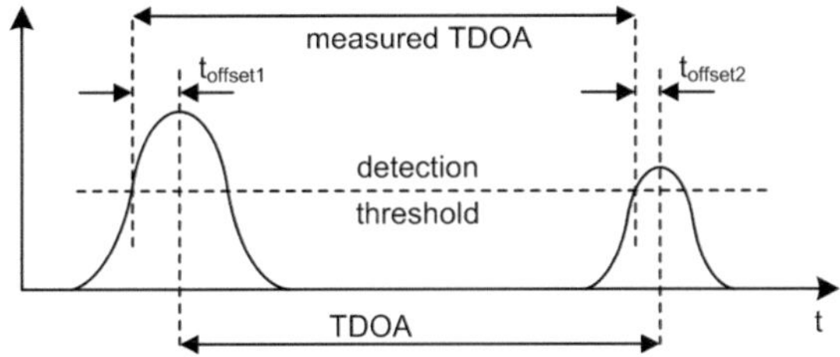

Fig. 4.15.: Trigger time dependency on threshold level and signal amplitude. At constant threshold level strong signals lead to larger time offset with regard to the true pulse maximum.

A detailed analysis of the threshold level influence on the trigger offset and, hence, the localization performance is given in section 6.2.1. Possible compensation methods are also named there.

Digital impulse stretching

The rectangular digital pulses at the comparator output have constant amplitude, however, their width (below 1 ns) is still proportional to the amplitude of the received pulse (the time when the pulse is above the threshold). Thus, they are not yet suited for direct further digital processing. In order to eliminate this effect, the differential outputs of the comparator are connected to a circuit responsible for extending the impulse to a constant length (cf. Fig 4.14) ([11],[Kol11]). The feedback loop controlling the resulting pulse width is connected to the clock-input of the fast D-flip-flop. When the logical "1" occurs at its input, the flip-flop generates a "high" signal. This is split between the output (e.g. connected to an FPGA) and the delay loop constructed of switchable inverters. When the signal reaches the flip-flop again it is reset and the "low" signal appears at the Q-output. The number of logic gates in the delay loop regulates the latency and thus the resulting pulse length. The constructed pulse stretcher can extend the incoming sub-nanosecond pulses to a duration of 3 ns to 10 ns. The realized circuit is depicted in Fig. 4.16(f). The narrower the output signal is, the higher the data-rate that can be received without causing the digital pulses to overlap.

After selecting the receiver architectures that will be compared and describing all their hardware components, their simulation- and measurement-based comparison will be presented in the next section.

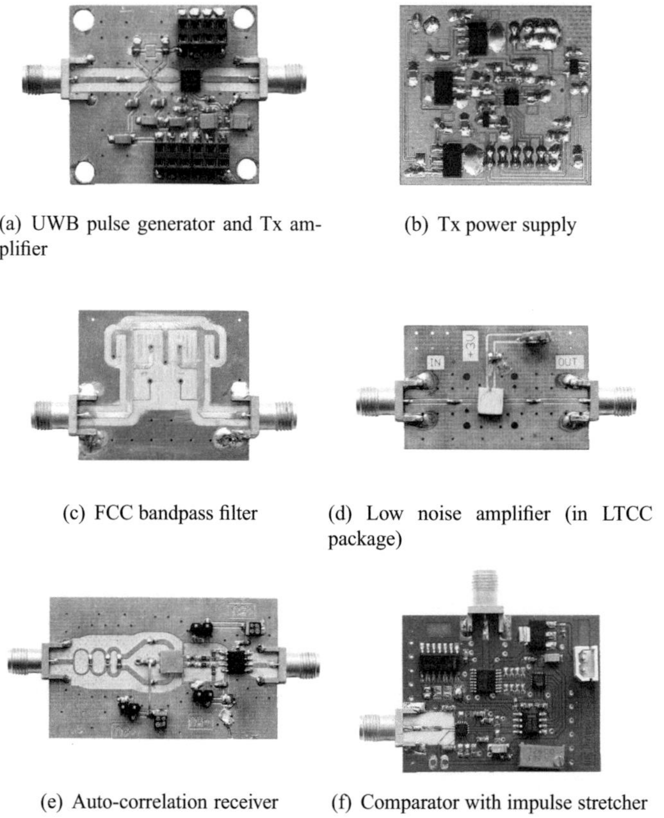

(a) UWB pulse generator and Tx amplifier

(b) Tx power supply

(c) FCC bandpass filter

(d) Low noise amplifier (in LTCC package)

(e) Auto-correlation receiver

(f) Comparator with impulse stretcher

Fig. 4.16.: Photographs of the realized hardware components, used during the measurements.

4.4. Performance comparison of the selected receiver types

This section begins with a simplified mathematical model of the baseband signal behavior of the three above-described UWB receivers. Furthermore, the description of the simulation model that has been used for benchmarking (best case conditions) is given. This is followed by the description of the measurement setup used for the performance verification. In the final part, the results are thoroughly discussed.

4.4.1. Mathematical description of the received signal in baseband

A simplified mathematical description of the signal behavior after the multiplier stage (correlator) in the three selected receiver architectures is given below. The following notation is used: the received signal, after propagation through the radio channel, amplification and filtering, is denoted as $r(t)$. This corresponds to the left column in Fig. 4.17 and consists of the useful signal $s(t)$ (distorted and attenuated impulse) and noise $n(t)$. The locally generated template in the CR (identical with the transmitted signal) is denoted with $e(t)$.

The analysis will be accompanied by references to the graphical representation in Fig. 4.17, showing two periods of a pulse train transmitted at 100 MHz PRF.

Using the above notation, the signal at the output of the multiplier in the CR can be written as follows:

$$r_{CR} = r(t) \cdot e(t-\tau) = (s(t) + n(t)) \cdot e(t-\tau)$$
$$= s(t)e(t-\tau) + n(t)e(t-\tau). \tag{4.1}$$

The optimal synchronization (overlapping of $s(t)$ and $e(t)$) is given, when the relative shift between those signals is $\tau=0$. In this case, the multiplication will result in the maximum pulse correlation, where the noise (before and after the pulse) is multiplied with zero (Fig. 4.17(a), middle). Otherwise the template misses the pulse and generates an output signal, consisting of the cross-correlation with noise and cross-talk (XT). The two products from (4.1) cannot occur at the same time, hence the plus sign between them does not mean an addition of signal amplitudes in physical sense. It much rather indicates the number of possible states at correlator output.

In the ACR the input signal is squared (Fig. 4.17(b), middle), resulting in only positive output.

$$r_{ACR} = r(t) \cdot r(t) = s^2(t) + 2\,s(t)n(t) + n^2(t). \tag{4.2}$$

The desired useful signal is the first element in (4.2). The squared noise is however unwanted, as it is not zero-mean anymore, hence cannot be removed by the adjacent low-pass filtering (DC offset in Fig. 4.17(b), right). The STR is thus degraded. The middle term in (4.2) represents the XT of the signal from both inputs to the output, as well as the cross-correlation with noise. As there is no time shift between the two multiplier inputs, this component contributes to the useful signal level and is not separately observable.

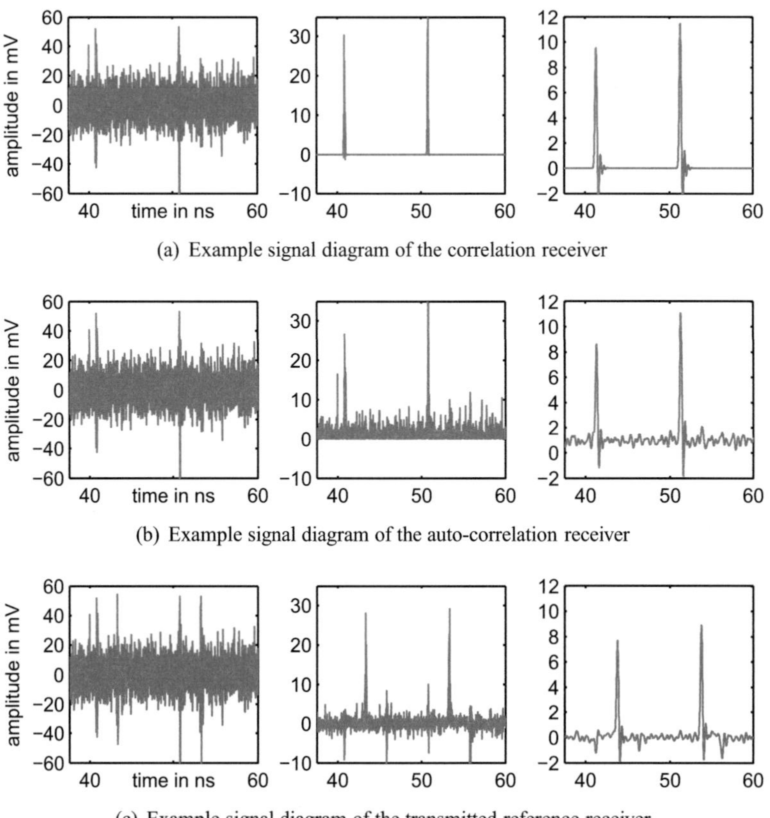

(a) Example signal diagram of the correlation receiver

(b) Example signal diagram of the auto-correlation receiver

(c) Example signal diagram of the transmitted reference receiver

Fig. 4.17.: Schematic representation of the received UWB pulse for different receiver types. From left to right: RF signal at multiplier input (after LNA and filter), behind the multiplier and after low-pass filtering.

As previously mentioned in section 4.2.3, the TR concept is meant to combine the template correlation advantages (high STR) of the CR with the simple architecture of the ACR. To avoid the synchronization issues, the template pulse is transmitted shortly after or before the actual data-pulse (delay τ_{TR}). This can be observed in Fig. 4.17(c), left: two additional spikes, following the pulses in 2.5 ns distance. The multiplication of the shifted signals results in:

$$r_{TR} = r(t) \cdot r(t - \tau_{TR}) \tag{4.3}$$
$$= s(t)n(t - \tau_{TR}) + s(t)s(t - \tau_{TR}) + s(t - \tau_{TR})n(t) + n(t)n(t - \tau_{TR}).$$

The second product in (4.3) is the desired correlation result of the two pulses (two highest spikes in Fig. 4.17(c), middle. The first and third term are responsible for XT and cross-correlation with noise, surrounding the useful signal at $\pm\tau_{TR}$ and visible in the same figure. The STR benefit compared to the CR originates from the fact that the last term in (4.3) is the cross-correlation of two shifted copies of the same noise signal. By definition they are uncorrelated and generate only zero-mean noise (assuming AWGN), that can mostly be filtered out (Fig. 4.17(c), right). Additionally, according to this notation the ACR can be considered a special case of the TR concept, with $\tau_{TR}=0$.

4.4.2. Simulation and measurement scenario

Software-based STR investigation
For the purpose of simulation-based STR investigation the three above described UWB transceiver structures: correlation receiver (Fig. 4.5(a)), auto-correlation receiver (Fig. 4.4(b)) and transmitted-reference receiver (Fig. 4.5(b)), have been implemented in the Agilent Advanced Design System Ptolemy [Agi09] simulation environment:

- the variable attenuator, bandpass filter, LNA and the power divider, have been implemented in the simulation ([10], [SDST07]) using the measured S-parameters,

- the waveform of the pulse generator has been captured with the oscilloscope and imported in the simulation,

- the correlator was modeled as a perfect multiplier, i.e. without I/O cross-talk,

- the antennas were modeled as gain blocks, as only one angular direction was used in the measurements (cf. section 6.3). The radio channel has been substituted by the attenuator, with attenuation proportional to distance (calculated for the center frequency f_M =6.85 GHz of the pulse),

- the time-domain simulation step was set to 10 ps, in order to achieve high accuracy,

- at each simulated PRF (see section 4.4.4) at a given distance the Tx amplitudes were identical for each transceiver type.

The simulation results will be further used as a benchmark.

Implementation of receiver architectures and measurement setup

In order to allow a fair comparison between transceivers, all architectures have been implemented using the same hardware components described previously in sections 4.3.1 and 4.3.2. The implementation aspects, specific for each structure are listed here:

- CR: The correlation receiver requires synchronous triggering of the template pulse, in order to achieve time alignment with the received pulse and hence the maximum output amplitude. In the experiment the template pulse was triggered synchronously with the Tx pulse, where the fine time alignment has been obtained using precise delay control IC (see section 5.3.1). The delay variations of this element, caused by its thermal drift, were partly stabilized by a heat sink.

- ACR: The auto-correlation configuration was achieved by inserting the power divider after the LNA and connecting its ports to both correlator inputs using 10 cm long coax cables.

- TR: The transmitted reference transceiver sent two pulses per one period of the trigger signal (cf. Fig. 4.2). The delay between both pulses has been set to $\tau_{TR} \approx 2$ ns. The delay in the transmitter and the receiver was realized using coaxial cables (true time delay).

The receivers were placed at 3 m distance to the transmitter and tested consecutively. The STR investigation by means of measurements was performed in a corridor, since in this environment the multipath contributions were negligible. The received signal was recorded after the correlator using a 12 GHz bandwidth oscilloscope at 40 GS/s. The oscilloscope was triggered synchronously with the Tx pulses. The test setup and the scenario are shown in Fig. 4.18.

To investigate the STR performance as a function of the Tx-Rx separation, a variable attenuator (cf. section 4.3.1) has been used before the antenna at the Tx site (cf. Fig. 4.1). This attenuation, added to the free space path loss, emulated the increase of the Tx-Rx distance. At every Tx-Rx separation, for every receiver type and PRF, 300 pulses were recorded. Seven different attenuation settings D_{set} have been used in the experiment. The resulting Tx

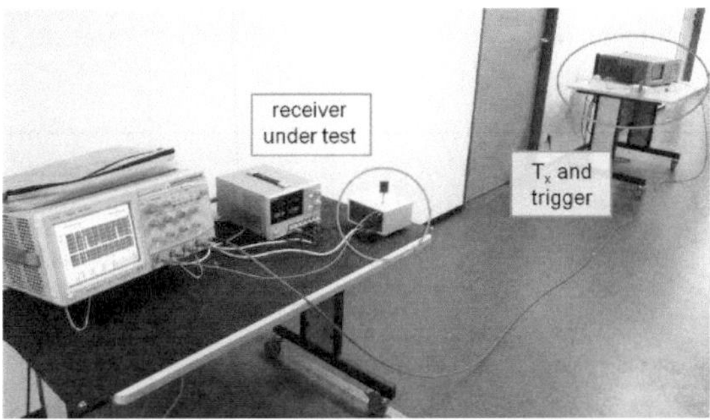

Fig. 4.18.: STR measurement setup. UWB transmitter and receiver under test at 3 m
distance. In the case of the CR, the cable at the bottom on the right side has
been used for template synchronization.

impulse amplitude A_{pp}, equivalent path loss PL and corresponding distance
R (determined for the pulse center frequency) are listed in table 4.1.

Table 4.1.: Transmit pulse peak-to-peak amplitude A_{pp} and the equivalent range R.

V_{pp} in mV	D_{set} in dB	PL in dB	R in m
900	0	58.7	3
600	3.6	62.3	4.54
435	6.6	65.3	6.41
293	9.6	68.3	9.06
235	11.6	70.3	11.41
186	13.6	72.3	14.36
150	15.6	74.3	18.08

4.4.3. Determining the STR based on captured time signal

This section gives an example how to extract the signal and noise statistics,
described in section 2.3.2, from the captured (simulated or measured) signal
and calculate the STR according to equation (2.23).

A measured output signal of the CR, presented in Fig. 4.19, will be considered
first. In the signal in Fig. 4.19(a), the received pulses are present (in the
case of the CR an optimal synchronization is required). The second signal
contains only noise (and, in the case of the CR, the cross-talk of the template
pulse). From several hundred measured pulses it is possible to find the average
peak level of the received pulse A_{peak} (dash-dotted line in Fig. 4.19(a)). The

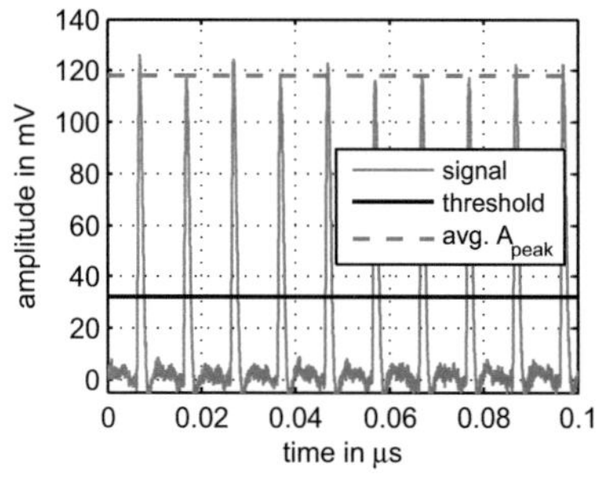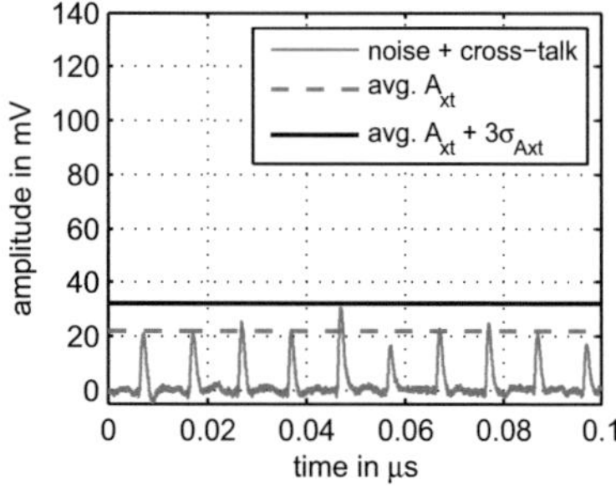

(a) CR output with optimal synchronization   (b) XT of the template pulses in the CR

Fig. 4.19.: Measured output of the CR: for an optimal alignment of the received signal and the template pulse (a) and if no pulses are received (b). The levels of the average correlation result A_{peak} and the estimated threshold are marked with a dashed and a solid line (a). The average template pulse cross-talk plus noise level A_{xt} is marked with a dashed line (b).

statistical parameters of the noise signal (average cross-talk level A_{xt} and its standard deviation σ_{Axt}) can be determined as in Fig. 4.19(b). Inserting those terms into equation (2.23) results in an STR. By definition the STR of 0 dB describes the case in which the received impulse is at the threshold level. Below this point no reliable reception is possible.

The calculation procedure for the ACR and the TR systems is very similar, with the exception that the noise and XT terms cannot be directly determined by turning off the useful signal, as they are coupled. The information about noise and cross-talk is obtained by removing the useful signal spikes in the post-processing and evaluating the remaining components. An exact description is provided in [Ber11].

4.4.4. STR-based performance evaluation

The simulation and measurement results for all investigated receiver types are presented in Fig. 4.20(a),(c),(e) and 4.20(b),(d),(f), respectively. Three different values of PRF have been investigated during measurement and simulation: 10 MHz, 100 MHz and 200 MHz. The principle shape of the curves resulting from simulation and measurement coincide well. In general, the measured STR is lower, which is caused by the implementation loss and the fact that in the simulation such elements as coaxial connectors and cables have not been included.

STR at distances below 6 m

The largest discrepancies between simulation and measurement occur at distances below 6 m and they range from 2 dB for TR and ACR, upto 4 dB for CR. The main reason for this effect are the relatively large pulse amplitudes at these distances. Hence, the simplified modeling of the analog correlator in the simulation does not account for the non-linear behavior (e.g. the saturation), nor the cross-talk of the pulse towards output. Those two effects contribute to noise during the measurements, reducing the STR.

High signal powers, and hence cross-talk, come in handy for the ACR, as in this architecture the cross-talk has nearly no influence on the noise level at the output; rather the opposite: it can contribute to signal power. As a consequence, above 4 m distance the difference to simulation for the ACR architecture is less than 1.5 dB. This characteristic feature of the ACR allows it to exhibit better performance than the TR at distances up to 4 m both in simulation and measurements. The Tx-Rx distance at which these curves intersect depends on the transmission power: the higher the Tx amplitude the larger the range.

At distances of 4 m to 6 m all measured curves run very closely. In this region the TR system seems to even outperform the CR. The reason for this is the previously mentioned small margin for the allowed synchronization error of the CR (overlap with the template pulse) [15]. The settings of the delay element [Mic11], used to adjust the trigger time of the template pulse, are susceptible to thermal drift. Due to these slight changes in the time delay the auto-correlation function does not reach its theoretical maximum and the CR performance degrades.

STR at distances up to 10 m

In this range the simulated curves exhibit the behavior suspected from the theory, where the TR architecture performs better than ACR, but worse than CR. Towards higher distances (or equivalent lower Tx powers) the differences between simulation and measurements become smaller. This discrepancy for the TR architecture is marginal above 6 m distance. Also the ACR curve shows excellent agreement between simulated and measured results, however in general the performance of this architecture start to degrade faster than others.

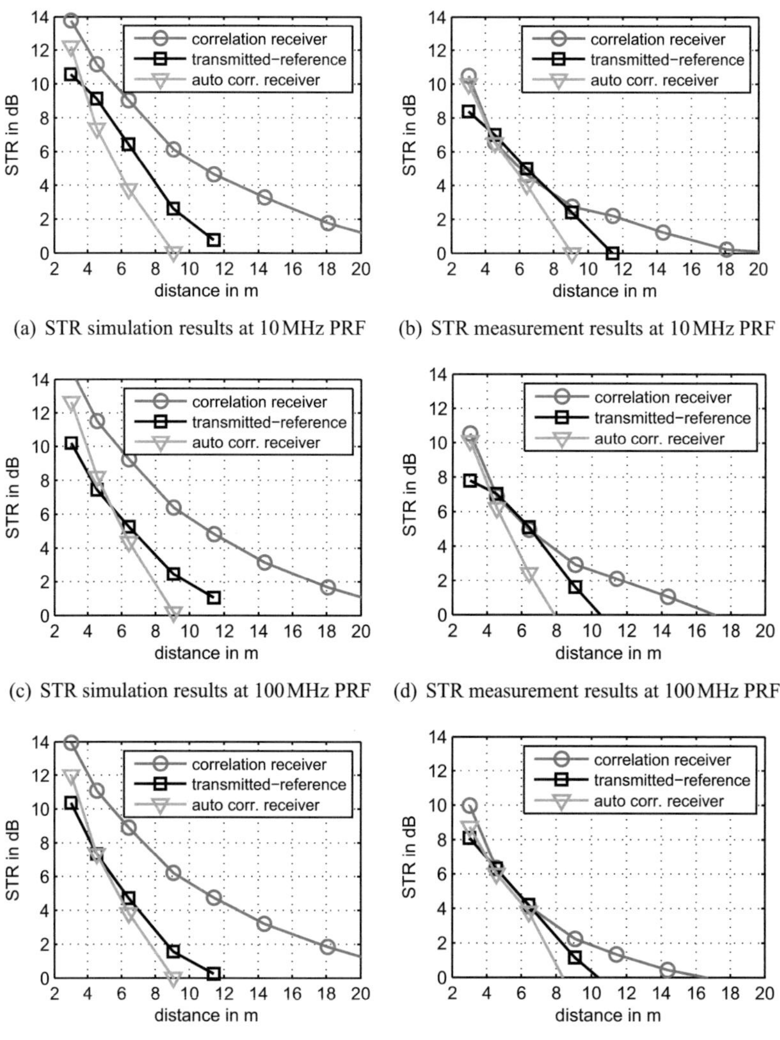

(a) STR simulation results at 10 MHz PRF

(b) STR measurement results at 10 MHz PRF

(c) STR simulation results at 100 MHz PRF

(d) STR measurement results at 100 MHz PRF

(e) STR simulation results at 200 MHz PRF

(f) STR measurement results at 200 MHz PRF

Fig. 4.20.: Signal-to-threshold ratio as a function of Tx-Rx separation at various transmission rates.

The measured TR and CR curves still run very closely, making it once again clear how important is the perfect synchronization for the performance of the correlation receiver.

STR at distances above 10 m

At these distances (Tx powers) the ACR cannot operate, neither in simulation nor in practice. The operation range of the TR ends shortly above 10 m, wheres the difference between simulation and measurement is less than 1 m. The CR can still operate reliably, while the STR curve runs ≈ 2 dB below the simulated one. Here the correlation gain becomes prominent and the characteristic falls much less linearly (in dB-scale) than in case of the ACR and the TR.

Conclusions and closing remarks on receiver evaluation

The performed investigations show that, in general, the theoretical advantage of the correlation receiver is not prominent at small distances. In measurements, the CR can significantly outperform the other two architectures first at larger distances (above ≈ 9 m). However, this advantage is only present if an ideal synchronization is assured. Even a time misalignment of ± 15 ps between the received and the template pulse can severely degrade the STR performance.

The auto-correlation receiver is a good choice, if the architecture should be kept simple and the signal to noise ratio at the receiver input is high (either for the very short ranges or the high transmission power).

The transmitted-reference receiver offers a good trade-off between the complexity and achievable STR, which is synonymous to a large transmission range. This advantage comes however at a cost of the poor transmit power efficiency of the TR, as compared to CR and ACR twice the amount of pulses (energy) per useful bit are required.

The shapes of the curves in Fig. 4.20 remain almost identical with changing PRF. This demonstrates the correctness of the performance evaluation based on instantaneous power approach for the SNR calculation in impulse-radio systems. In practice, there is no simple classification of the UWB receivers, with regard to their SNR performance. The intersection points of the STR curves indicate that there is no clear best solution with respect to operation ranges. Thus, depending on the application in mind, a trade-off has to be made. The low receiver noise figure and the careful selection of the low-pass filter bandwidth can directly improve the STR of all architectures.

As the correlation receiver has shown the best performance in terms of the achievable distance, it will be examined in a more detailed way. In the next chapter, the advantages and drawbacks as well as the practical aspects of the CR implementation will be thoroughly investigated.

5. Synchronization of the UWB correlation receiver

According to the results presented in the previous chapter, the most effective method of UWB pulse reception in presence of AWGN noise is to correlate the received signal with a template pulse (TP). As mentioned in section 4.2.3, the correlation can be performed either with a clean template that is locally generated in the receiver, or a dirty template. The dirty template is distorted in order to account for the influence of the channel and therefore can improve the correlation result. However, the full advantage of the correlation receiver can be taken only under the condition, that the synchronization between the template and the received pulse is optimal, which is seldom the case in practice. Because of this, the emphasis of this chapter is put on the following aspects of CR operation:

- the required pulse synchronization accuracy and methods of achieving it,

- synchronization and pulse tracking algorithms, as well as their practical realization, and

- the efficiency of this concept in terms of transmission stability, under realistic operation conditions.

This chapter is structured as follows: First, the complexity of the UWB receivers synchronization issue is outlined in section 5.1. Here, the theory of template synchronization and tracking methods are described. In section 5.2, the most important parameters of the CR subsystems are introduced, along with the associated requirements. Following that, in section 5.3, the hardware realization of the CR is described together with the measurement setup and the verification results of the previously proposed synchronization schemes. Furthermore, in section 5.4, the experiment showing the potential strength of the CR is presented. Finally, the proposed approaches and the realized system are compared in section 5.5 with the solutions encountered in the literature.

5.1. Template synchronization and tracking

The synchronization process of the correlation receiver is inherently associated with data communication. As the emphasis of this study is on the localization aspects, the "dummy" data consisting of a constant pulse train will be used when testing the CR synchronization schemes. This however can easily be extended to transmission of proper data by introducing pulse modulation scheme (e.g. pulse position or amplitude). This aspect is briefly described in section 5.1.2 and tested in section 5.3.3.

When using a pulse train, the term *data rate* can be represented by the pulse repetition frequency PRF_{Tx} of the transmitter, or the period of the signal $T_{PRF\,Tx}$. In order to demodulate the received signal with the same data rate, the relative time shift between the transmitter and the receiver has to be equalized. This will ensure a high output signal level from the correlator, hence a high probability of the pulse detection.

The UWB pulses are short and the duty cycle (ratio of the pulse width to the transmission period) is typically below 1 %. To perform the correlation, the received and locally generated pulses need to overlap. Fig. 5.1 (left) shows the received pulse (black) and the TP (gray, dashed) and Fig. 5.1 (right) depicts the auto-correlation function (ACF) of the received pulse. Depending on the detection threshold (dash-dotted line), the receiver can synchronize at the main or side lobes of the ACF. For UWB pulses with a bandwidth of several GHz, a timing accuracy of several tens of picoseconds is required [CZ07]. For pulses used in this work (see. Fig. 4.7(b)) the synchronization margin is only 30 ps [1].

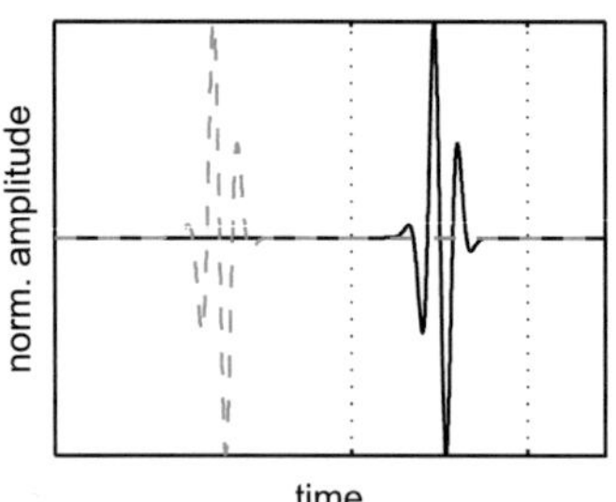

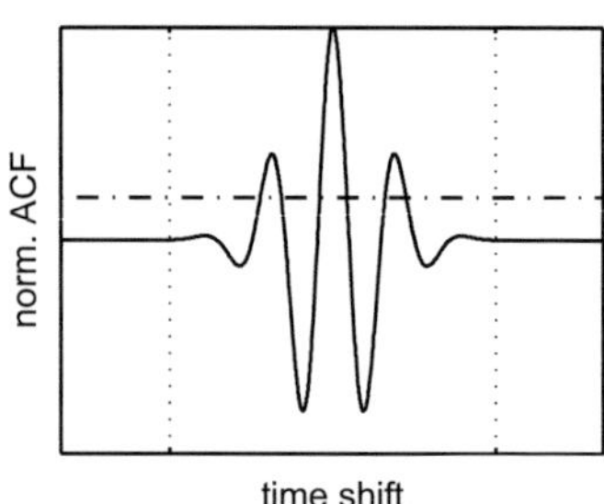

Fig. 5.1.: Two pulses (left): template pulse (gray, dashed) and received (black); (right): the ACF that occurs as the template sweeps along the received pulse (twice the pulse width); horizontal dashed-doted line represents the detection threshold level.

The synchronization problem in general is the question of precise control of the template pulse. The synchronization methods can be divided into two groups, with regard to the way in which the TP trigger signal is controlled.

Frequency-difference method

In the frequency difference method, the two UWB pulses (one at the transmitter and one at the receiver) are triggered with slightly different periods $T_{PRF} = 1/PRF$ [LSG$^+$09]. This is represented in Fig. 5.2 (top and middle). After each period the pulses approach each other with a time step $\Delta\tau$, until they finally overlap.

$$\Delta\tau = |T_{PRF\,Rx} - T_{PRF\,Tx}| \qquad (5.1)$$

In the worst case, when the pulses are close to overlap however the search is performed in the opposite direction, this process can last up to a maximum duration of:

$$t_{search} = \frac{(T_{PRF\,Rx})^2}{\Delta\tau} - T_{PRF\,Rx}. \qquad (5.2)$$

At the moment when the two pulses overlap, the period duration of the receiver template signal is switched to the one of the transmitter. Hence, all the following pulses will overlap, as long as the transmitter or receiver do not move. Any relative movement between those two will result in a change of the pulse time of flight, hence a shift of the received signal versus template. This approach may seem simple in theory. It has however several drawbacks in terms of its practical realization, which will be discussed at the end of this section.

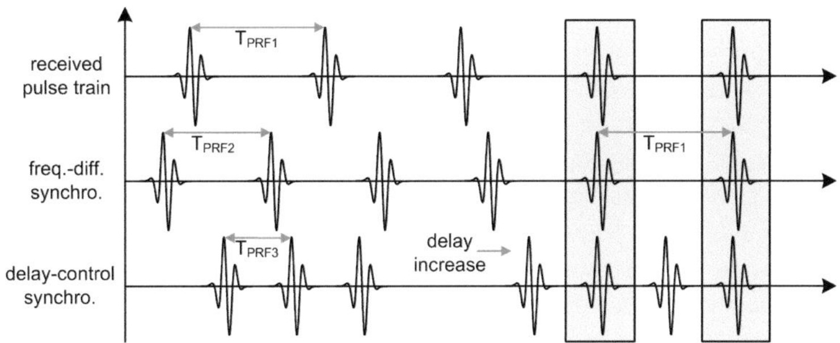

Fig. 5.2.: Operation principle of the frequency-difference (middle) and the delay-control synchronization method (bottom). The received pulse train is depicted at the top.

Delay- or phase-control method

The delay method follows a different approach from the frequency difference method. Here, the TP is triggered with an integral multiple N of the PRF_{Tx}, hence N pulses fit within one period of the transmitted signal (see Fig. 5.2(top and bottom)). This repetition rate remains constant during the entire process. In order to achieve the shift in the time domain, the templates are delayed stepwise in the receiver. While triggering N template pulses (one Tx period) the correlation result is being observed. If no pulse was detected, the delay of the TP trigger signal is changed and another full period is awaited. The delay adjustment can also be interpreted as the phase change of the TP trigger signal. This allows for a N-fold acceleration compared to the duration derived in equation (5.2). As soon as the received pulse and the TP overlap, the currently used delay will not be changed further. In this case, the detection occurs at every N-th template pulse.

The major advantage of the delay method is that no change in the PRF_{Rx} is needed. In this approach it is required to change the delay in possibly small steps in order not to overleap the received UWB pulse.

Comparison of both methods and discussion

From the practical point of view, the realization of the frequency-difference method is connected with very high effort. Firstly, there exists an unavoidable signal latency in the receiver (between the correlator and the decision block that sets the TP PRF) due to the physical size of the circuit. This latency is in the order of magnitude of several, up to more than ten nanoseconds. Due to the reaction time of this control loop, the change of the PRF by the analog clock source (e.g. a VCO) is delayed. Meanwhile, the TP will still be triggered with the previous un-updated PRF and the synchronization will be most probably lost. This effect can be compensated by calibrating the reaction times and accounting for them while changing the PRF. However, the temperature-dependent behavior of the control loop elements would have to be very well stabilized, to ensure that the changes of trigger signal period are much smaller than the required synchronization accuracy of 30 ps.

Secondly, the PRF in the receiver has to be changed very rapidly, when the pulses start to overlap. The analog clock sources such as e.g. voltage control-led oscillators (VCO), would reach their performance limit. This is caused by the contradiction between achieving a low jitter trigger signal and the speed, with which the PLL can change between two PRFs.

Provided that the reaction time of the receiver remains below the transmitter period duration, and that the time to the generation of the next TP is precisely calculable, the above problems can be bypassed by employing a digital generation of the template trigger signal.

Because of those reasons, the delay-control approach was identified as the one that is more suitable and has been implemented in this study. When integral multiples of the transmitter PRF are used for triggering the TP, the switching of the delay element is not a time-critical operation. However, in the presence of clock discrepancies between the transmitter and receiver (see section 5.3.3), the calculated relative shift according to equation (5.1) has to be compensated using tracking (see section 5.1.2).

5.1.1. Synchronization schemes

The time between the two template pulses will henceforth be called the *search window* (SW). Below, the three methods of sampling the search window with the template pulse are discussed. The number of degrees of freedom in choosing the parameter values for each of those methods is strictly limited by the pulse shape used and the available hardware. The performance of those methods, depending on the choice of parameters is described in section 5.3.2.

Linear synchronization (LS)
During LS, the entire search window is sampled with an increasing delay, as shown in Fig. 5.3(a). The decision as to whether or not the synchronization has been obtained is based on several observed pulses at the same position. The parameters that can be adjusted are the delay step (time span between two closest sampling points) and the amount of observed pulses at one delay setting.

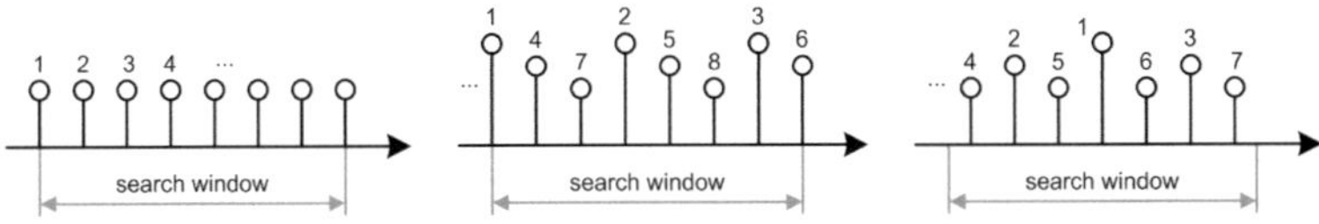

(a) Linear synchronization.   (b) Linear-offset synchronization.   (c) Tree synchronization.

Fig. 5.3.: Three investigated synchronization schemes. Horizontal axis represents time, wheres the circles represent the trigger time-points of the template pulse.

Linear-offset synchronization (LOS)
Here, in contrast to the LS, the delay is increased by one fine step first after the entire SW was sampled with coarse increments; then another sweep is conducted. This is shown in Fig. 5.3(b). This method allows a faster synchronization than the LS in the case when the ACF is broad, i.e. in the case when the signal is strong. If this is not the case, the LOS will still sample all positions.

Tree synchronization (TS)
In the previous two synchronization schemes the search starts at the beginning of the SW and the absolute delay increases monotonously until the end of the SW. Using TS the SW is equally sampled with decreasing step widths. This is shown in Fig. 5.3(c). This method represents a width search of a "tree" (also called *breadth-first search*).

Under the assumption that the smallest delay step fits n-times in the SW, the TS search begins in the middle of the window at $n/2$. In the following step, the interval is bisected and the template signal is placed in the center of the left subinterval, at $n/4$ and then in the middle of the right subinterval at $3n/4$. This is the second level of the algorithm. In the following levels, the delay steps will be further reduced until finally the smallest available value is reached. After each delay change the correlation result is assessed. To avoid multiple sampling of the same positions, the delays of the lower level lie always between the ones of the upper one. This approach should result in a faster synchronization in the cases where the main pulse is surrounded by many side lobes (e.g. originating from close multipaths).

5.1.2. Tracking phase

After the synchronization phase is completed, the template pulse is triggered in a way to meet the received pulse exactly at the maximum of the ACF. However, from this moment on, any change of the pulse time of flight (e.g. due to relative Tx-Rx movement) or a deviation in the pulse periods will lead to a loss of synchronization and increase of BER. To prevent this from happening, a tracking mechanism capable of following the changes in the pulse arrival times is required. In order to determine the changes in the radio channel the pilot pulses (PP) are used, while transmitting the data. In this section, the pulse position modulation (PPM) scheme is considered, where the data-information is coded in the time separation between one PP and two possible positions of the following data pulse (DP).

The idea of tracking is to follow the leap between the low and high pulse detection probability (equivalent to low and high BER). This leap will be referred to in the following as an *edge*. It can be distinguished between two versions of this procedure, called in following the single- and double-edge tracking.

Single-edge tracking

In the single-edge tracking (SET) the goal is to follow the leading edge of the ACF. After the successful synchronization phase the position of the TP is precisely known and from here on the tracking begins. In case when the pulse was detected at a certain TP delay setting, in the next period the TP trigger delay is reduced. This procedure continues until the PP is not detected any more. Then the delay is increased until the pulse is detected again. In this way the TP will be continuously switching between two states: pulse detected and pulse undetected. This is depicted in the top part of Fig. 5.4(a). The gray rectangle represents the width of the ACF, where the pulse can be detected, whereas the circle indicates the trigger position of the TP. The arrow indicates the direction the TP will be shifted in the next cycle.

In the case when the ACF edge changes its position, it will be tracked without affecting the transmission of actual data. In case when the relative shift of the Tx signal would be increasing too fast (compared to the time interval between two PPs) the SET will not be able to follow and the transmission will be interrupted. This can happen, either due to the very large difference in the Tx-Rx clock cycles (cf. section 5.2.1), or PPs laying too far apart. The maximal difference of the pulse periods at the transmitter and receiver that is still compensable by the SET depends on the number of PPs per time unit. The schematic representation of the situation where the TP can barely follow the PP is depicted in the lower part of Fig. 5.4(a). The relative movement of the Tx-Rx pulses is represented by the arrow in the gray area (shift of the ACF).

During SET, the delay required to detect the DP is calculated from the TP offset to the last successful detection of the PP. Depending on the direction in which the PP-template was moved compared to the last period, one extra delay step is either added or subtracted from the DP-template.

Double-edge tracking

During double-edge tracking (DET), in contrast to SET, both edges of the ACF are observed and evaluated. The PP is alternately sampled at the early-edge (white circle) and the late-edge of the ACF (black circle). The beginning of the tracking is analogous to the one of SET. The delay of the early-TP is

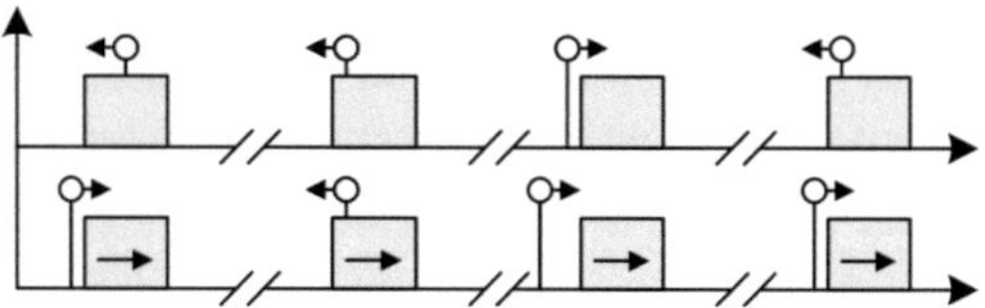

(a) Tracking of the pilot pulse at one edge - SET.

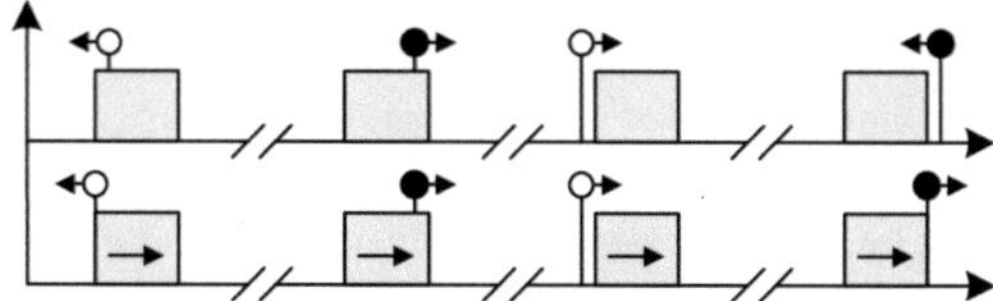

(b) Tracking of the pilot pulse at two edges - DET.

Fig. 5.4.: Two tracking schemes used for following the pilot pulse. For both schemes two cases are considered: the ACF (gray rectangle) is static (top) and when ACF exhibits time drift (horizontal arrow within the gray rectangle) while being observed over several Tx periods (bottom).

successively reduced until the ACF edge is found. Meanwhile, the late-TP is moved stepwise towards the late-edge. This is shown in the top part of Fig. 5.4(b).

The key advantage of DET compared to SET is the possibility to precisely determine the position of the ACF maximum, which is the difference of the late and early delay divided by two. This allows the placement of the DP-template in the optimal position and minimize the BER. An additional advantage of this technique is that even when the relative time shift is increasing at a high rate, one of the TPs (early or late one) will most probably be able to detect the PP.

5.2. Precise digital clock signal sources

In this section, a brief overview of the methods suitable for generating very precise trigger signals is given. First, two important parameters which characterize the signal sources are defined. Then, possible implementations are discussed.

5.2.1. Precision clock sources: figures of merit

Time jitter (TJ)
Jitter is the time domain representation of phase noise in the frequency domain. In the case of digital signals, jitter describes how much the rising or falling slopes of the signal differ from their ideal position. In coherent UWB transmission, the jitter of the template signal must be much smaller than the width of the auto-correlation function.

Frequency accuracy (FA)
The FA of an oscillator is the offset from the specified target frequency and is expressed in *parts per million* (ppm). The synchronization quality depends on the oscillator's FA due to the fact that the time offset is the integral of the frequency error. The larger the difference between the PRFs at transmitter and receiver, the more effort is needed to maintain the synchronization. Off-the-shelf-oscillators offer an accuracy at a level of $\pm 5\,$ppm to $\pm 100\,$ppm.

5.2.2. Precision clock sources: possible implementations

In this section a brief overview of possible precision clock implementation is given. The focus lies here on their limitations in terms of frequency accuracy and jitter performance.

Phase Locked Loop (PLL)
One possibility to trigger the template pulse is with an oscillator integrated in a PLL. The frequency accuracy of the PLL is determined by divider used in the feedback loop; this allows only coarse frequency settings. TJ is dependent on the lowpass filter: a narrow filter for low TJ does not allow for fast frequency changes without the PLL losing lock. This makes this concept only conditionally usable for the envisioned system, since it would require large T_{PRF} and minimal PRF shift between Tx and Rx, hence slow synchronization.

Direct Digital Synthesis (DDS)
The average FA of the DDS depends on the TJ of the reference oscillator, divided by the number of bits of the frequency tuning word. The average FA in the range of nHz is achievable [Van00]. If the length of the control signal from the phase accumulator is longer (number of bits) than the size of the phase register, this leads to cropping, and hence approximating of the phase values at the output. As a result unwanted phase-jitter appears in the output signal [Che11].

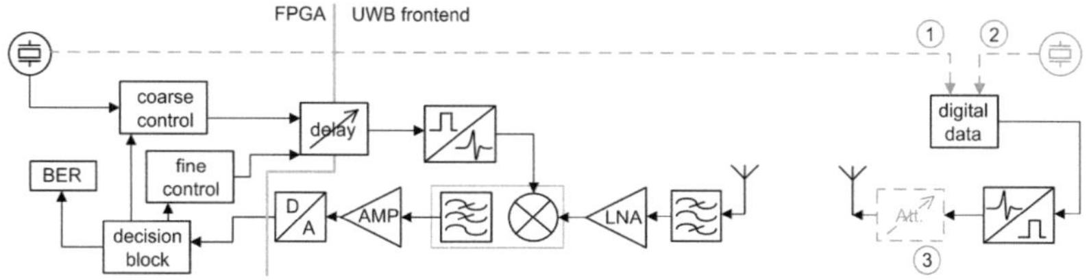

Fig. 5.5.: System configuration used for verification of synchronization and tracking algorithms. The gray dashed elements 1, 2 and 3, as well as wires are optional.

Serial Shift Register (SSR)

The accuracy of the output digital signal depends on the clock frequency of the register. To achieve the accuracy of 30 ps the SSR would need to be triggered at 33 GHz, which is not reasonable in the envisioned application. This method however can be still used for coarse synchronization, e.g. to shift the TP through the SW in 1 ns steps (1 GHz clock required).

Programmable Delay (PD)

PD allows to precisely adjust the phase of the digital signal. The delay step size of the devices available on the market reaches down to 10 ps [Mic11]. In conjunction with e.g. a PLL, the PD can be used for very accurate frequency or phase adjustments [TYM+06].

After the analysis of possible alternatives and feasibility of their realization, the combination of programmable delay element (characterized in section 5.3.1) and the phased locked loop was identified as the most suited for the project and has been implemented.

5.3. Verification of the correlation receiver

This section starts with the description of the hardware used to verify the algorithms presented earlier in this chapter. Furthermore, a parametric study of the synchronization and tracking algorithms is presented and discussed.

5.3.1. System setup for evaluation of synchronization algorithms

The schematic representation of the measurement setup is depicted in Fig. 5.5. The field-programmable gate array (FPGA) supplies the clock signal (PLL with adjacent PD), which directly triggers the impulse generator. After an optional attenuation the pulse is radiated by an antenna and propagates towards the receiver. After reception, the signal is bandpass-filtered, amplified[9] and the pulse is multiplied with the locally generated template. The correlation result is 1-bit A/D converted. The comparator's output is acquired by the FPGA. Based on that, the decision block (implemented synchronization scheme) manage both the coarse and the fine clock control blocks. The coarse control (SSR) allows for a timing precision of 1 ns and the fine one adjusts the programmable delay in 10 ps steps (10 bit delay control).

In Fig. 5.6 the characteristic of the delay IC used is presented, which indicates that the delay bins are not spaced exactly in 10 ps increments. The RMS delay setting error is 2.35 ps and is strongly temperature dependent.

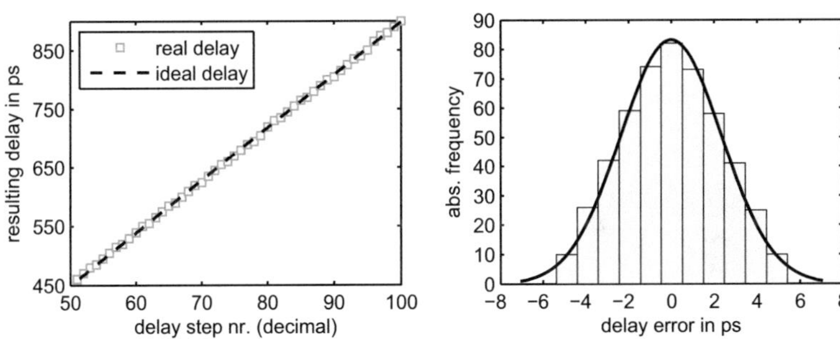

Fig. 5.6.: Characteristic of the used SY89295U delay IC: comparison of target and actual delay (left) and the absolute frequency of setting errors quantized in 1 ps steps (right). For the purpose of the plot clarity only 50 states are presented in the left figure. The right histogram involves 512 states.

For UWB signal generation, the pulser circuit from section 4.3.1 was used, emitting 5th-derivative Gaussian pulses. The ACF, schematically shown in Fig. 5.1(right), indicates that in order to remain in the region of at least 80 % of the maximum, a timing tolerance of only 30 ps is allowed [1]. The rate

---

[9]Although this order yields worse noise figure, it is necessary to filter the potential out-of-band interferences before amplification, since the employed LNA exhibits a very low 1 dB compression point.

at which the pulses are transmitted is 12.5 MHz (period of 80 ns). The TS is triggered with 200 MHz (5 ns period) to accelerate the search (N=16, cf. section 5.1). The Tx-Rx separation is 2.5 m.

With this setup, it is possible to evaluate several parameters. If the synchronization path (1) is used and the clock (2) is turned off, the synchronization algorithms can be tested without including other effects (e.g. no FA issues). With (2) on and (1) off, the sensitivity of the algorithms against FA of both clocks can be tested. Moreover, a variable attenuator (3) allows the modification of the SNR in a controlled way. The scenario with the deployed system is depicted in Fig. 5.7.

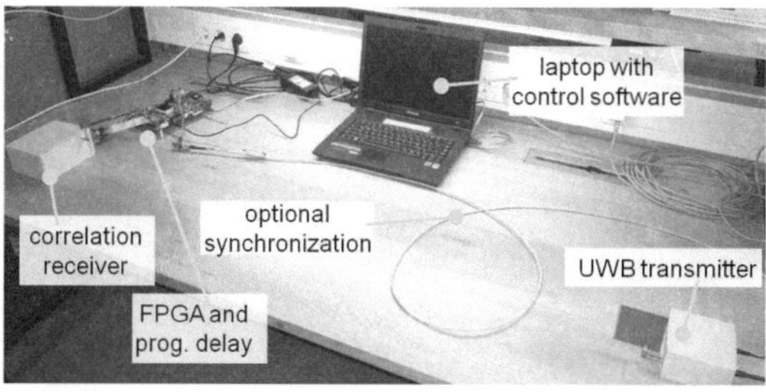

Fig. 5.7.: Photograph of the correlation receiver measurement scenario.

5.3.2. Performance comparison of synchronization schemes

In this section, the different synchronization algorithms, described in section 5.1.1, are evaluated and compared with each other. Within each synchronization scheme the characteristic parameters are investigated. First, the evaluation criteria are defined:

- The time duration until the synchronization is finished: this is the number of transmitted pulses, with a period of 80 ns that has elapsed until the synchronization has been completed.

- The synchronization threshold: this is a ratio between the number of pulse periods (transmitter), for which time the system will remain at the same delay setting (receiver), and the number of pulses that need to be detected in this time in order to declare synchronization as completed.

For test purposes, a real indoor environment was chosen, where the multipath propagation is very prominent. The transmit signal amplitude was set to 950 mV (peak-to-peak). The result is a large number of correlation side lobes in received signal originating from multiple reflections and a low main correlation maximum.

Linear synchronization accuracy at different threshold settings
Firstly, the received signal after propagating through the previously described scenario, is sampled with the CR. The delay element controlling the template impulse is adjusted in a way that at every setting 10000 pulses are observed. This is continued to cover the entire SW and is represented in Fig. 5.8 (SW=5 ns). The delay setting is on the abscissa and the ordinate indicates the number of received bits at this position in percent (probability of detection). In this graph, a large number of correlation side lobes is present.

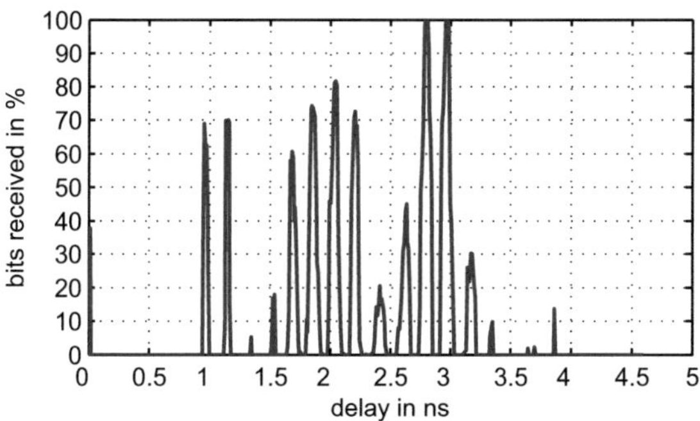

Fig. 5.8.: Measured correlation result of the template pulse with the received one. The 5 ns represent the width of the search window.

In this measurement however, the absolute number of pulses observed during the synchronization process, hence also its duration, was fixed. This can be optimized by using relative detection thresholds. In order to investigate this better, the LS scheme was chosen and several parameters were varied. In the case when 1000 incoming pulses are to be observed at one delay position and the criterion for synchronization completion is 990 received pulses, this results in a relative detection threshold of 99 %. If only 100 pulses are to be observed at one delay setting, and after 99 received pulses the synchronization is declared to be completed, theoretically this will result in the same threshold, but a shorter time required for synchronization.

In Fig. 5.9 (top), a cut-out of the section with the highest percentage of received bits from Fig. 5.8 is shown (between 1.8 ns and 3.1 ns). To investigate different thresholds, the LS with a 10 ps step is used. The results are shown in Fig. 5.9 (bottom). The lower threshold of only 100 observed pulses (circle marker) shows a very large variance. The reason for this is that in the correlation side lobes it is possible to detect 99 pulses at a stretch. When the observation period of 1000 pulses is used, the synchronization is more reliable and tends to complete the process at the same position (small variance). A longer observation time at each position assures better statistical certainty. With increasing threshold the search result is closer to the main maximum (2950 ps). In most cases however, the synchronization ends at the strong side lobe (2800 ps). Only in two cases, using a threshold of 99.9% (triangle marker), the true position at 2950 ps was found.

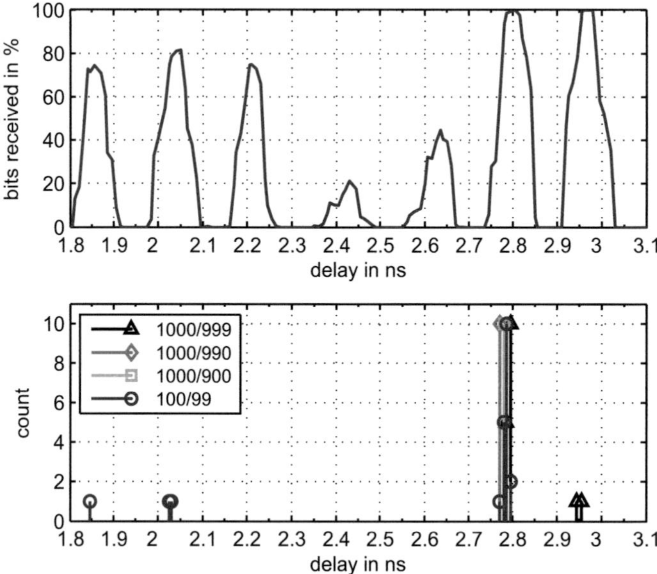

Fig. 5.9.: Comparison of the synchronization thresholds (observed/required): zoom-in of the measured correlation function from Fig. 5.8 (top) and the delay where the stop criterion was met (bottom). The main peak position is 2950 ps, the others are the correlation side lobes and echoes.

Regarding the threshold choice and the number of pulses to be observed, a trade-off has to be made. A large amount of pulses, hence a high threshold, will lead to a more accurate and reliable synchronization. On the other hand, when the Tx and Rx modules are operating with physically separate oscillators (signal path (2) in Fig. 5.5), a certain clock drift will be present. In this case, during the long observation time at one position the searched pulse will migrate, leading to even longer search times. In the worst case, synchronization will not be possible at all. As a consequence, for the tests described in the following two paragraphs the absolute threshold of 500 pulses received in a row was used.

Influence of the step width in the linear synchronization scheme
The results of the LS test for different delay step widths are presented in Fig 5.10. Note that a new channel measurement has been taken. The search duration (ordinate) expressed in Tx signal periods is plotted against the determined delay (abscisse). The squares are the values for the smallest delay step of 10 ps. In this case, more time is required for synchronization than if a delay step of 20 ps is used (rhombi). Moreover, the values of the determined delay using the 10 ps LS scheme exhibit considerably larger variance that the 20 ps version. Both however often synchronize on the side lobes of the ACF that are spaced approx. 150 ps from the maximum. The results of the synchronization performed with 30 ps LS are marked with circles. Here, only every third delay setting is used, which results in the fastest synchronization. This version consistently (only one outlier) synchronizes on the proper delay. These tests were performed at a distance of 3 m.

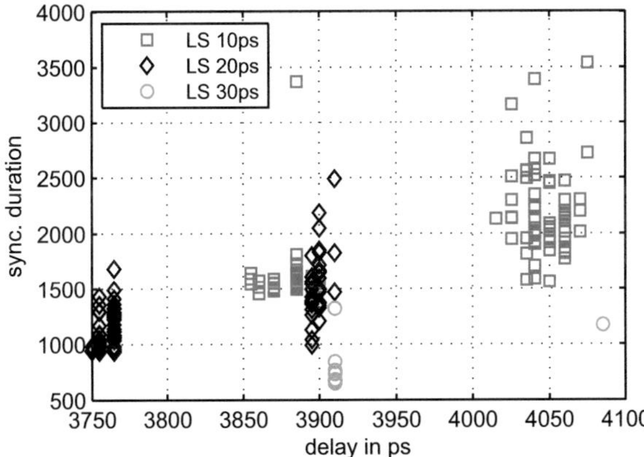

Fig. 5.10.: Behavior of the linear synchronization at various step widths. The determined delay on the abscissa is plotted versus synchronization duration expressed in Tx periods. The true position of the ACF maximum is 3920 ps.

Comparison of synchronization schemes
Here, the LS, LOS and TS schemes are evaluated under the same conditions and compared with each other. This data has been obtained as previously at a distance of 3 m and each synchronization scheme was tested 100 times.

In Fig. 5.11, the LS with a step width of 30 ps (circles, the same as in Fig. 5.10) is compared with LOS (rhombi). Both schemes perform well and regularly synchronize on the correct delay. Moreover, the values lie close to one another with the exception of one outlier of the LS. The TS scheme (squares) delivers similar results as the LS with a 20 ps step from Fig. 5.10, however requires more time to achieve this.

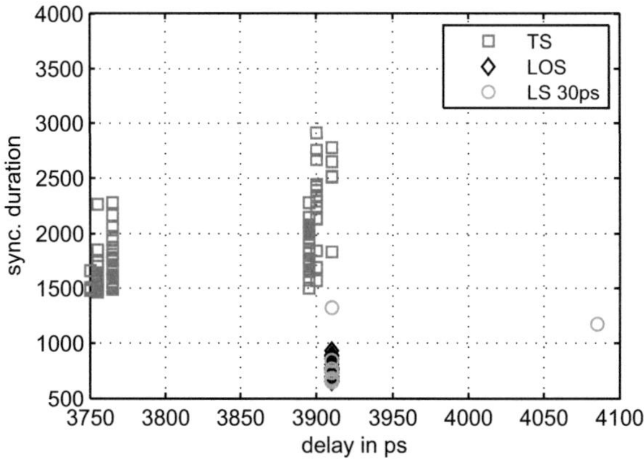

Fig. 5.11.: Behavior of the linear synchronization versus linear-offset and tree synchronization. The determined delay on the abscissa is plotted versus synchronization duration expressed in Tx periods. The true position of the ACF maximum is 3920 ps.

Synchronization performance in dependence of SNR

In the further test, the variable attenuator was used to influence the transmit amplitude (option (3) in Fig. 5.5). The attenuation was changed in 2 dB steps between 0 and 10 dB. At each setting, all the above presented measurements were re-performed and the average values for each scheme (determined delay and required time) are presented in table 5.1.

Table 5.1.: Synchronization effort of LS, LOS and TS schemes averaged over different SNR.

scheme	delay in ps	duration in μs	times aborted
LS 10 ps	3920.5	142	21
LS 20 ps	3961.8	142.5	132
LS 30 ps	3914.3	134.5	151
LOS 30 ps	3916.2	108.3	151
TS	3887.8	142.8	100

The finest linear synchronization offers the best results at the cost of a relatively high search time. Although larger delay steps of the LS slightly reduce the search time, they often do not allow reliable synchronization and the synchronization process is aborted. With LOS it is possible to obtain synchronization after the shortest time. The number of situations where no synchronization was possible in LOS is the same as in the case of LS with 30 ps step. In this test the TS and 20 ps LS exhibited the worst performance among the investigated algorithms. They not only required most time to synchronize, but also the found delay lies ≈ 40 ps away from the maximum.

5.3.3. Synchronization and tracking of an autonomous transmitter

The assumption that the transmitter and the receiver can operate using the same frequency source is often not realistic. Therefore, the autonomous operation of both devices will be investigated in this section. For this test, separate clocks were used for the transmitter and the receiver (option (2) in Fig. 5.5). Due to the limited frequency accuracy of both sources, the obtained synchronization will be lost after short time. For example: for a relative FA of 30 ppm and transmitting at 12.5 MHz PRF this would take only $\approx 1\,\mu s$. Therefore, the decision threshold (number of pulses observed at one delay setting) has to be investigated. In the previous section it was shown that a longer observation (higher threshold) yields better results when the clocks are synchronous. In the case of the asynchronous operation, high thresholds will not give the desired result. On the other hand, setting a lower threshold increases the probability that the receiver will synchronize at a side lobe of the ACF or even at a multipath echo.

In the conducted experiment, the transmitter clock was measured to be running 0.754 ps slower per each period (80 ns) of the receives' one. This means that within 40 cycles the received pulse will shift against the template by more than 30 ps, hence losing the synchronization (cf. section 5.3.1). In the first phase, the time (i.e. the number of Tx cycles) required to obtain the synchronization was measured as a function of the detection threshold. The results are presented in Fig. 5.12.

Here, the LS method with 10 ps step width was used for synchronization and the measurements were performed seven times at each threshold setting. It is evident, that the linear increase of threshold causes an almost logarithmic increase in the search time. The average values of the synchronization duration shown in Fig. 5.12 are summarized in table 5.2. Based on those

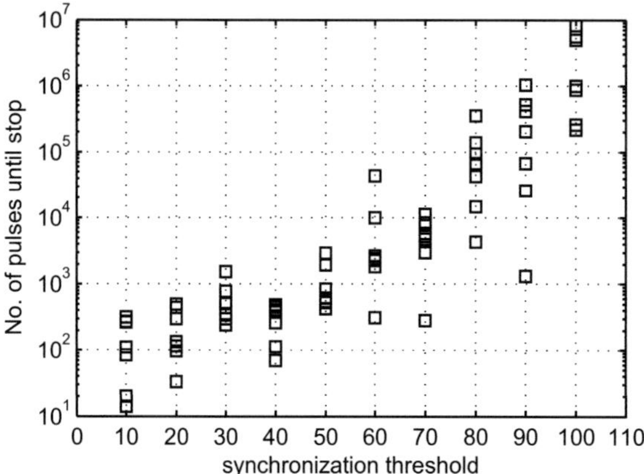

Fig. 5.12.: Unsynchronized clocks: duration of the synchronization process as a function of threshold (pulses required to be detected in a row).

results it becomes obvious that in order to achieve synchronization within an acceptable time interval the threshold has to be set very low, compared to the results obtained in section 5.3.2. This however, lowers the probability of synchronization to a maximum of the ACF. This fact can have its repercussions in the quality of the adjacent tracking phase.

Table 5.2.: Average synchronization duration at different threshold settings.

threshold	10	20	30	40	50	60	70	80
duration in cycles	155	231	604	309	975	9260	5752	101660
duration in µs	12.4	18.5	48	24.7	78	740	460	9600

	90	100
	177580	1983100
	14200	158000

In the second part of the above experiment, the quality of the tracking was investigated. After successful synchronization (threshold set to 50 out of 55 detected pulses) the receiver immediately switched to the tracking mode (SET, cf. section 5.1.2) and attempted to remain synchronized for the following 300 bits (bit = pilot + symbol pulse). The experiment was repeated 50 times for each of the three different synchronization steps of LS. The results are presented in Fig. 5.13.

110

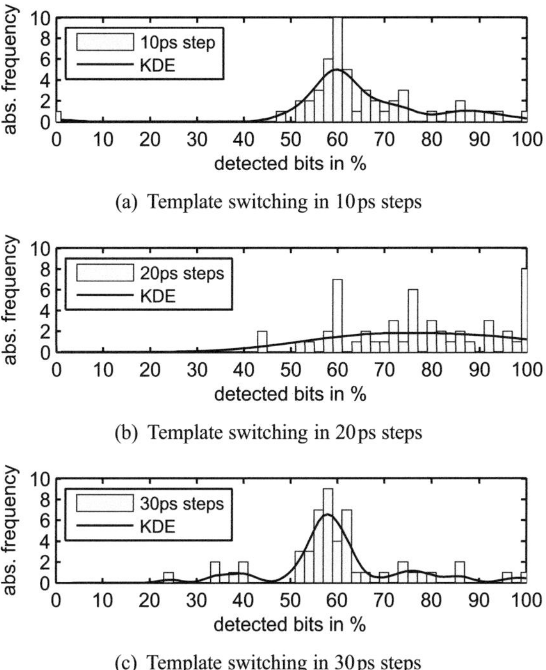

(a) Template switching in 10 ps steps

(b) Template switching in 20 ps steps

(c) Template switching in 30 ps steps

Fig. 5.13.: Tracking of the signal from the asynchronous transmitter, using different template delay steps with regard to the last known position of the pilot pulse.

The histograms show the percentage of correctly detected bits and how often (i.e. absolute frequency) this result was achieved. The solid line represents the kernel density estimate (KDE[10]), however, from this amount of samples, no definitive conclusion regarding the statistical distribution can be drawn. As demonstrated in Fig. 5.13, it is possible to achieve a high percentage of detected bits in the tracking phase. This requires however a careful choice of the the step width with which the pilot pulse is followed by the template, which depends on the relative FA of both clocks. For the here designed system and parameters used, the 10 ps step is too small and 30 ps step too large. The best tracking results have been achieved with a 20 ps step, where in average $\approx 75\%$ bits were detected.

---

[10]KDE is used for non-parametric estimation of the probability density function of a random variable, when using a limited amount of data samples.

One of the factors contributing to the lower tracking reliability is the situation when the delay exceeds the width of the SW and the delay element needs to be reset to zero to allow further tracking. As the delay element employed exhibits strong temperature dependency, the reset point may change and in consequence the pilot pulse is lost [Mic11].

5.3.4. Conclusions regarding the data communication using correlation receiver

The synchronization schemes introduced have been implemented on a Xilinx Virtex 5 FPGA and tested on an FCC compliant transceiver (cf. previous chapters). A parametric study was performed to identify the most efficient algorithm. In terms of search time, the LOS algorithm is the best one, however, the LS with 10 ps has the best accuracy. Tests performed with a separate clock at the transmitter and receiver indicate that the synchronization and tracking are very sensitive to frequency accuracy and can only be established for a short time period. In practice, this kind of a system exhibits a very high complexity. The advantage of a better STR (compared to ACR and TR) does not fully justify the effort and increase of the raw BER as demonstrated above. This method can however be used in radar or ranging applications, as the accurate synchronization is then directly related to distance accuracy. An experimental example is given in the next section.

5.4. Application of correlation receiver for precise distance measurements

One of the possible applications of the CR and the related synchronization algorithms is the precise range measurement. In this configuration the programmable delay element is the key component since it determines the synchronization quality, and hence the ranging accuracy. The PD described in section 5.3.1 exhibits a RMS delay accuracy of 2.35 ps, what corresponds to ≈ 0.7 mm distance accuracy. To investigate the ranging performance of the system introduced in the previous section, the following measurement setup has been used:

- the Tx and Rx modules were connected as in Fig. 5.5 (signal path (1)),
- the UWB transmitter was placed on the linear positioner (similar to Fig. 6.7)

- and the correlation receiver was positioned in 1.5 m distance from the Tx.

- The Tx was moved away from the Rx in 1 mm steps and covered a total distance of 20 cm.

At each position, the TP was swept through the entire SW using the LS scheme with 10 ps steps, whereas at each delay step 10000 Tx pulses were evaluated to obtain the ACF. From the maximum of the ACF, the time of signal arrival is determined and can be related to the distance by which the Tx has been moved (relative to an arbitrary selected starting position). The results are presented in Fig. 5.14.

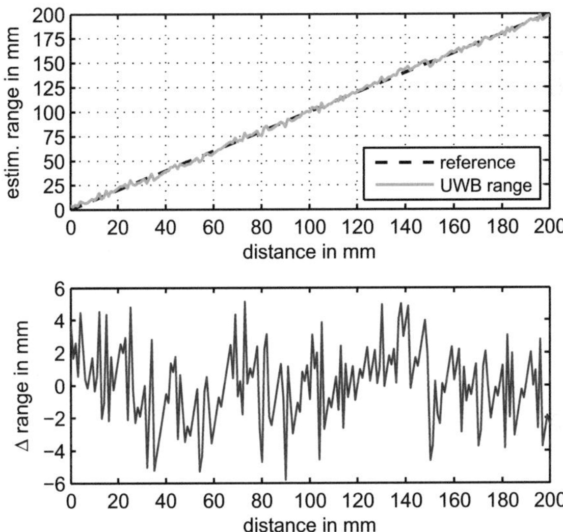

Fig. 5.14.: Results of the ranging experiment with the correlation receiver over 20 cm distance (top) and the resulting +/- range error (bottom).

The distance estimated by the CR-ranging setup follows the true one very accurately (reference marked with dashed line in Fig. 5.14). The range residues, depicted in Fig. 5.14 (bottom), originate from the discrete nature of the TP trigger signal (PD+PLL). The calculated RMS range error is 2.28 mm and corresponds to ≈ 7.5 ps. Hence, the distance measurement error is three times larger than the PD accuracy alone. The accuracy could be further improved by curve-fitting to approximate the measured ACF and using the maximum of this function instead of evaluating the TOA at which the most pulses were received.

5.5. Comparison with other experimental systems

There are very few implementations of the IR Ultra-Wideband correlation receiver that can be found in the literature. The system designed in this study is compared with the two best ones published up to date, and the parameters are summarized in table 5.3.

Table 5.3.: Comparison of the realized CR performance with literature.

source	signal BW / duration	CLK gen. method	min. time step	possible schemes	temp. PRF	ranging precision	raw data rate
[BHY$^+$02]	2 GHz (0.5 ns)	PLL	60 ps	all but TS	10 MHz	—	10 Mbps
[TYM$^+$06]	500 MHz (2 ns)	PD	160 ps	all above	—	±2.5 cm	1 Mbps
this work	7 GHz ~ 250 ps	PLL +PD	10 ps	all above	200 MHz	±2.3 mm	12.5 Mbps

A PLL-based CR was implemented in [BHY$^+$02]. The use of the VCO, controlled through a DAC allowed a minimum step of 60 ps. Compared with the 10 ps step presented here, it does not allow for an accurate synchronization in the case of such narrow, higher order derivative pulses. Moreover, the PLL requires a certain time to settle and does not allow erratic phase changes, therefore, no TS synchronization is possible. In addition, the implementation presented in this study allows a TP triggering at a rate 20-times faster than in [BHY$^+$02], significantly reducing the time required for synchronization.

In [TYM$^+$06], the PD approach was chosen. The UWB pulses had only a 500 MHz bandwidth, compared to 7 GHz in this work, and a delay step of 160 ps was possible. The raw data rate (equivalent to PRF) was 1 Mbps, compared with 12.5 Mbps implemented here, and the maximal ranging accuracy was ±2.5 cm, whereas here the ±2.28 mm accuracy has been demonstrated. Additionally, the operating range of the here designed system is several meters, whereas in the solution presented in [TYM$^+$06] is only ≈ 1 m.

5.6. Conclusions and final remarks regarding correlation receiver

In this chapter the three possible synchronization and two tracking methods, for use in impulse-radio correlation receiver, have been introduced and parameterised. For the first time in literature all these schemes were implemented on a CR realized in hardware and extensively tested. The results indicate that:

- The fastest synchronization is offered by the linear-offset method, however, the 10 ps linear synchronization is the most accurate one.

- During the operation with separate clocks in the transmitter and receiver the frequency accuracy is the most critical system parameter, forcing the use of low detection thresholds.

- During the tracking phase the setting accuracy of the programmable delay element is a limiting factor. The choice of delay step size for tracking depends on the relative frequency accuracy of Tx and Rx clocks.

This implementation outperforms other architectures presented in the literature in terms of speed, accuracy, bandwidth and transmission rate. Although in this study the pulse trains (dummy data) were used for ranging test, the designed architecture is also ready for communication applications, where only the pulse modulation scheme (e.g. PPM) needs to be implemented.

The presented ranging results are very promising, and the correlation receiver is a good candidate for TOA, two-way-ranging applications or elliptical localization where the synchronization is required anyway. It is however not the first choice for the "pure" hyperbolic TDOA localization system, as the time-difference architecture operates in an asynchronous manner. For this reason, the localization demonstrator, described in the next two chapters, has been realized based on auto-correlation receivers.

6. Localization system design and position error bound

In this chapter, the detailed architecture of the TDOA localization system is introduced and analyzed. The emphasis is put on the critical components, such as the time measurement unit and antennas, as they are jointly responsible for the achievable positioning accuracy and precision. In section 6.1, the top-level architecture of the developed TDOA localization system is explained. This is followed by a description of the time-difference measurement unit in section 6.2, including an origin analysis of a TDOA variation and time bias errors. Sections 6.3 and 6.4 cover the topic of suitable antenna choices and the methodology of compensating their influence on the localization accuracy. Finally, based on the investigated system parameters and developed models, the expression for the estimation of the localization error lower limit is derived in section 6.5 based on the Cramer-Rao bound.

6.1. TDOA localization system architecture

An example of the localization system layout based on the TDOA technique is shown in Fig. 6.1. The APs are interconnected to ensure synchronization and an autonomous MU equipped with an UWB tag transmitts the impulses. The signals propagate within the scenario over physically different paths (channels) to reach the AP. The received signals from all the APs are forwarded to the time measurement unit through the synchronization network. The received signals undergo certain delays ($T_{\text{stop N}}$) before they trigger the time measurement. This can be expressed with the following equation:

$$T_{\text{meas N}} = T_{\text{TOF N}} + T_{\text{AP N}} + T_{\text{stop N}}, \tag{6.1}$$

where $T_{\text{AP N}}$ stands for the AP specific delay, e.g. in the RF path (between antenna and ADC). The first received impulse triggers the time measurement and the relative differences to the following impulses are calculated. Such a system requires an initial calibration to determine the $T_{\text{AP N}}$ and $T_{\text{stop N}}$. This architecture is used in this work to demonstrate the localization capabilities

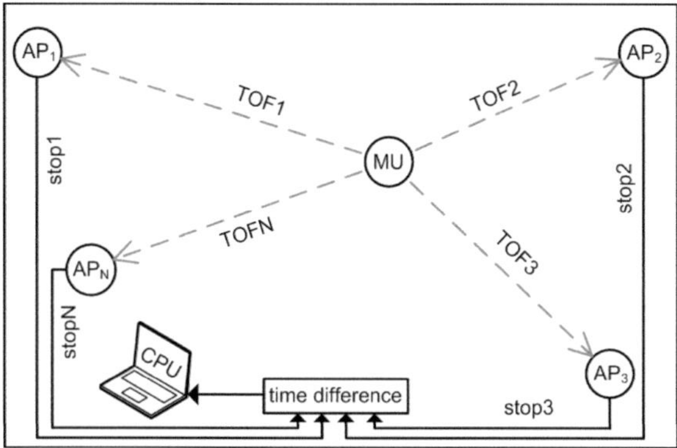

Fig. 6.1.: System configuration for TDOA localization with N synchronized access points and one mobile user.

of the designed UWB system and to investigate its performance limits. In the next section the influence of the time measurement unit on localization performance is investigated.

6.2. Precise time measurements

Precise time interval measurements are essential for the TDOA localization system and the precision of this action has a direct influence on the overall system performance. For this purpose, a device called the time-to-digital converter (TDC) can be used. It is used to measure the time elapsed between appearance of two (or more) signals at its input ports. The first incoming signal generates a *start*-event; the following ones, *stop*-events. The TDC used in this project is the ATMD-GPX produced by Acam® [aca07]. This model, depending on the operating mode, is capable of detecting two incoming digital signals with $\approx 25\,$ps resolution. The physical phenomenon that allows the achievement of such a high time resolution is the strictly determined pulse propagation time through logic gates in the TDC:

- the first incoming signal triggers the time measurement,

- the rectangular pulse starts to propagate through the logic gates, whose propagation times are precisely known,

- as soon as another incoming signal appears, the measurement is hold and the number of logic gates that the signal has passed through is read.

- The time difference between the two incoming signals is directly proportional to the number of logic gates that the pulse has traveled through between the start and stop events.

This methods allows only discrete time readings. The more advanced TDCs that can be found in the literature can reach a precision of 10 ps [JMK06]. Recently, a TDC with 1 ps precision was reported [KMK11].

In the following three sections, the most important parameters of the TDC are investigated and analyzed in terms of their influence on localization performance.

6.2.1. Threshold vs. trigger offset

The dependency of the threshold level, or, alternatively, the pulse amplitude, and the measured TDOA value is depicted in Fig. 4.15. The rising-edge-triggering, using one comparator threshold level (cf. section 4.3.3), leads to a premature, and hence, an incorrect time measurement. The higher the pulse amplitude, the more prominent the effect. Such behavior will potentially have an influence on the accuracy of the localization system.

To investigate this effect more thoroughly, the following measurement setup has been conceptualized: The UWB transmitter has been placed at a 1 m distance to the receiver (in ACR configuration) and triggered with a constant PRF. The detected baseband pulses have been synchronously recorded with the oscilloscope (averaging function set to 16 to reduce noise influence). The measurements were repeated for the different Tx signal attenuation levels, which emulated the increasing Tx-Rx distance (similar to section 4.4.2). During the off-line processing of the recorded data the time separation between the true pulse maximum and the point where the threshold was crossed has been evaluated (25 mV have been used as threshold). This separation is named *trigger offset* and its characteristic is presented in Fig. 6.2(a).

It is evident that at small distances the trigger offset is large and can reach values larger than 600 ps. However, for larger distances, where the pulse barely crosses the threshold, it decays towards zero. The operation of any impulse system in this range is however not practical, as this is a low STR region. This is visualized in Fig. 6.2(b), where the trigger offset is plotted against the corresponding ratio between pulse amplitude and threshold level: at ratio equal to 1, the STR equals 0 dB. Based on this measurement, the estimated trigger-offset changes linearly, at a rate of 487 ps per 10 m distance (pure path-

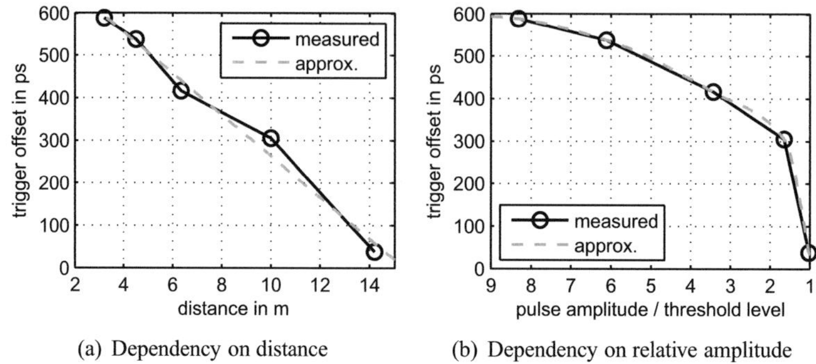

(a) Dependency on distance

(b) Dependency on relative amplitude

Fig. 6.2.: Trigger time offset from true maximum as a function of distance (a) and equivalently signal-to-threshold ratio (b).

loss, no antenna effects). The authors of [DDM$^+$09] also reported this effect and its linear characteristic, while analyzing a two-way ranging system.

To demonstrate the influence of this effect on the localization system accuracy, a simple scenario was investigated. Four access points have been deployed in the corners of a $10\,\text{m} \times 10\,\text{m}$ area and the time differences to the mobile user placed at $(1\,\text{m},1\,\text{m})$ have been calculated. One section of this area, showing the AP and MU, is presented in Fig. 6.3. The ideal TDOA hyperbolas are plotted with solid lines; they all cross exactly at the MU position marked with a rhombus. Furthermore, the TDOA calculation was performed, including the above described effect of trigger offset, caused by the respective MU-AP distances. As a result, the hyperbolas (dashed lines) shift towards the closest AP, as it has the strongest signal causing the early triggering, thus resulting in an illusive shortening of the distance. The cross marks the position of the MU, calculated based on the trigger-offset-error afflicted data. The resulting localization error is $7.8\,\text{cm}$.

Those considerations prove the necessity of a compensation of this effect in a localization system that aims at an accuracy below one decimeter. One of the possible correction methods is based on an iterative approach, where first the rough MU position is calculated based on the measured time differences. In the next step, the relative MU-AP distances are determined. Based on this and the information from Fig. 6.2(a), the original TDOA values can be corrected by removing the respective offset, resulting in a more accurate position estimation. In this approach, the antenna gains have to be taken into account, as not only the pure AP-MU distance is relevant, but the overall

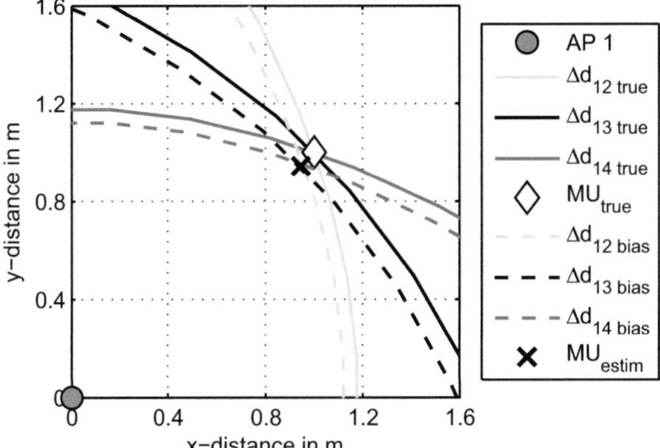

Fig. 6.3.: Time offset influence on the accuracy of TDOA positioning: visualization with hyperbolas.

path-loss. A similar method is explained in section 6.4.1 in greater detail. The results of its practical application are demonstrated in chapter 7.

6.2.2. Trigger time variance dependency on distance

Another important aspect that requires a closer investigation is the dependency of the trigger time variance on the Tx-Rx separation. This effect leads to a changing TDOA standard deviation, depending on the distance. Additionally it superimposes with the previously described trigger-offset error.

In order to analyze this phenomenon, the received pulse shape has to be considered first. The realistic form of the UWB receiver's correlator output (noise free) is depicted in Fig. 6.4 (thick solid black line). For triggering, only the rising pulse edge is considered, since the falling edge and ringing are not relevant here. The steepness and smoothness of the rising edge are essential for the reliable triggering, and hence the time measurement. The steepness of the slope can be expressed in V/s and is represented by a line tangential to the pulse edge at the point where it crosses the threshold. For an infinitely steep pulse edge, the angle that this tangent creates together with the threshold (α_{tangent}) will reach its maximum of 90°; for very week pulses, barely crossing the threshold, this angle will go towards 0°. The approximation of the rising pulse edge with the tangent is fully justified in this case, as the employed pulse has nearly ideal linear slope between 10% to 90% of its amplitude. In further considerations, the tangential angle of $45°$ is related to the rising edge steepness of $100\,\text{mV/ns}$.

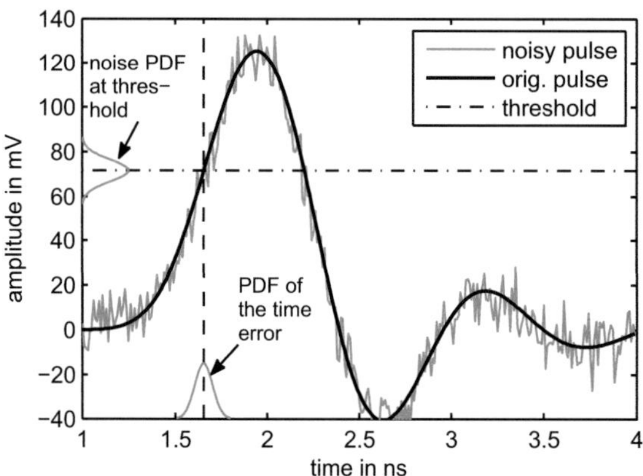

Fig. 6.4.: Relation between the amplitude variation of a noisy signal and the time measurement noise.

Now the noisy signal at the correlator output will be considered. It is represented in Fig. 6.4 by a thin gray solid line. The amplitude of the added noise is only for visualization purposes and is much higher than in reality. The rapid fluctuations of the signal amplitude, close to the threshold, can lead to a premature or late triggering. As all fast-changing signals have been already filtered by the low-pass filter placed at the output of the correlator, the only fast-paced voltage swings at this receiver stage can originate from thermal noise.

The root mean square of the thermal noise voltage, that is equal to the standard deviation, assuming the noise is zero-mean, is defined as:

$$v_{\text{n,RMS}} = \sqrt{4k_B T Z_S \Delta f}, \tag{6.2}$$

resulting in $\approx 82\,\mu\text{V}$ at the temperature of $300\,\text{K}$ (k_B denotes the Boltzmann constant), measured on $50\,\Omega$ load within $8\,\text{GHz}$ bandwidth (equivalent input bandwidth of the used comparator [Ana09]). The thermal noise is added to the output signal of the correlator. They both undergo an amplification by the baseband amplifier (voltage gain of 6.5; see section 4.3.2) before the threshold detection is done at the comparator. At this stage the $\sigma_{\text{noise}} \approx 530\,\mu\text{V}$. The statistical distribution of the noise fluctuation at the threshold point is

Gaussian[11] and is represented by the PDF projected on the amplitude-axis in Fig. 6.4.

In order to determine the influence of the signal level fluctuations on the time measurement statistics, σ_{noise} has to be divided by the slope steepness according to (6.3). E.g. for a 45° slope ($\approx 6\,m$ Tx-Rx distance from previous section), the resulting σ_{TDOA} is 5 ps. The resulting distribution in Fig. 6.4 is projected on the time-axis:

$$\sigma_{TDOA} = \frac{v_{n,RMS}}{tan(\alpha_{tangent}) \cdot steepness}. \tag{6.3}$$

In order to gain a better impression of the role this effect will play at a particular distance, the dependency of the tangent angle versus the distance was investigated. For this purpose, the pulses recorded with the measurement setup from section 6.2.1 were evaluated and the results are plotted in Fig. 6.5(a). It can be observed that for small distances the pulse slope is steep, becoming more flat for distances close to the maximum operational range. The equivalent representation related to the pulse amplitude is shown in Fig. 6.5(b). It indicates weak quadratic behavior, which coincides very well with the results presented in Fig. 4.20 (STR vs. distance).

An example calculation based on the above considerations indicates that for a distance of $\approx 4\,m$, hence a tangent angle of $\approx 60°$, the projection of the amplitude noise on the trigger time noise results in a time measurement uncertainty of $\sigma_{TDOA}=3\,ps$. At 14 m distance and a tangent angle of 12° the $\sigma_{TDOA}=25\,ps$. In order to obtain the overall time measurement σ, the above values need to be added to the intrinsic time measurement precision of the TDC. The jitter of the comparator (situated between the correlator and TDC) is at least one order of magnitude smaller than the trigger time variations caused by the thermal noise ($\approx 200\,fs$ RMS for the here used ADCMP572), and hence it is of no concern in this system.

These theoretical considerations were verified by measurement and the results are presented in Fig. 6.6. The measurement setup was here very similar to the one from Fig. 6.7(a). Here however the linear positioner was not used, and the Tx pulse was attenuated to emulate the distance change instead.

In Fig. 6.6(a), as predicted above, the rise of the time measurement variance can be observed with increasing distance. At lower distances, the TDC precision governs the overall time measurement precision, whereas above 10 m the noise begins to dominate the results. Below this distance the measured values

---

[11]thermal noise is approximately white, meaning it has nearly a Gaussian probability density function

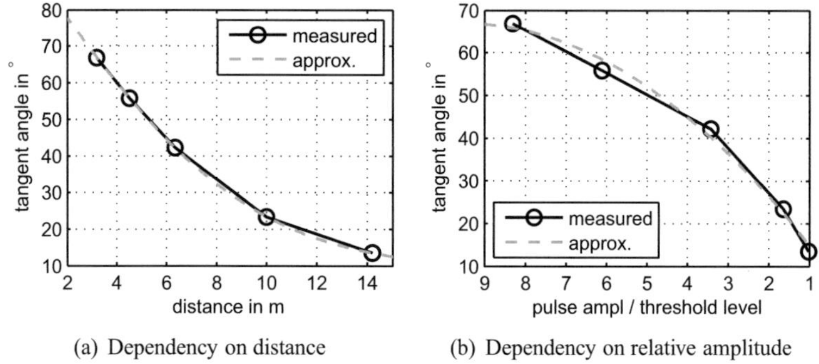

(a) Dependency on distance   (b) Dependency on relative amplitude

Fig. 6.5.: Tangent angle in the threshold point as a function of distance (a) and equivalently signal-to-threshold ratio (b).

follow the predicted ones very well. At higher distances, where the pulses are already weak, even a small error in determining the tangent angle results in a significant discrepancy of the true and predicted σ_{TDOA}. This is due to the $tan(\alpha_{tangent})$ dependency in equation (6.3).

Unlike the trigger offset, the variance of the time measurement cannot be directly compensated for. Assuming a zero-mean noise, the averaging method is an effective tool to reduce the variance [Kol11], and hence to obtain more precise TDOA measurements. The variance of the time measurement is, besides the MU-AP geometry, the key parameter determining the overall precision of the localization system (see equation (3.65)). The authors in [EQK$^+$12] attempt to give a closed formula expression that relates the σ_{TDOA} to the position error, however, this is only done for the 1D case and does not include any geometrical information.

6.2.3. Resolvable time difference and relative ranging

The interesting aspect of positioning is the distance measurement resolution and RMS accuracy that can be achieved with the TDC. For this purpose, the setup as in Fig. 6.7(a) was used, where the Tx and Rx were aligned in a straight line. The initial distance and the pulse amplitude were set to emulate the Tx-Rx separation of 6.3 m. The receiver was moved in 1 mm steps away from the transmitter and the TDOAs were measured. As presented in Fig. 6.6(b), the inherent resolution of the TDC is equal to 25 ps (bar spacing in the histogram). By evaluating the time-difference where the highest bar

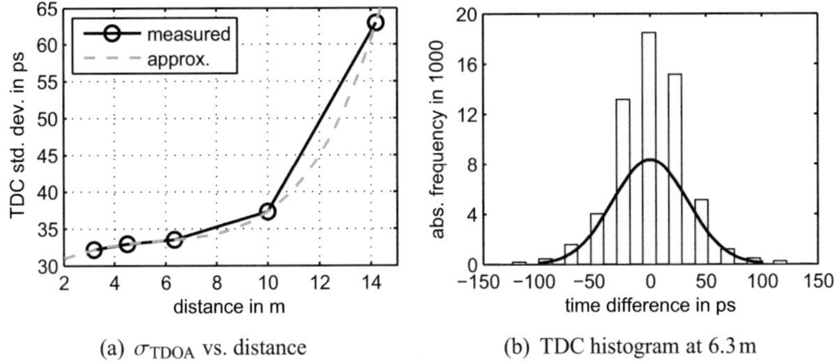

(a) σ_{TDOA} vs. distance

(b) TDC histogram at 6.3 m

Fig. 6.6.: On the left (a): the dependency of the TDOA standard deviation on the distance, influenced by σ_{TDC} and receiver noise. On the right (b): the histogram (absolute frequency) of the TDOA measurement at 6.3 m distance with a fitted normal distribution.

corresponding to the most often measured position is present, the distance could be measured only with discrete steps of 7.5 mm.

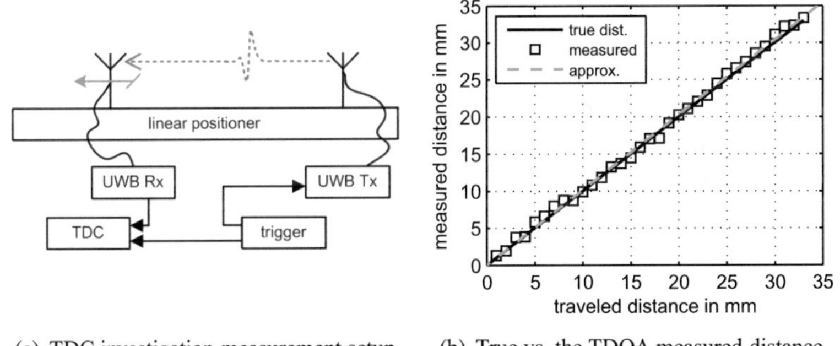

(a) TDC investigation measurement setup

(b) True vs. the TDOA measured distance

Fig. 6.7.: Measurement setup for investigation of the TDC performance in terms of relative accuracy and resolution.

However, by recording more values per position it is possible to fit the normal distribution and read the mean value. This was done to obtain Fig. 6.7(b), where a 33 mm long route along the linear positioner is shown. The true travel distance is plotted with the solid line and the measured relative distances, based on averaged TDOA values, are marked with squares. Additionally the linear fit curve of the estimated distances is plotted with a dashed line. The experiment shows that the measured distance increases slightly faster than the true one, achieving a value of 33.5 mm after 33 mm. This effect is caused by the trigger offset, described in section 6.2.1. As the trigger offset increases with $\approx 50\,ps/m$, this corresponds to 1.65 ps offset after 33 mm, and hence, exactly to the 0.5 mm difference between true and measured distance.

This proves that it is feasible to eliminate the trigger offset using the method discussed in section 6.2.1. After calibrating the trigger offset, the RMS distance estimation error using the TDC was 0.46 mm. This shows that by employing a sufficient amount of TDOA averaging and offset estimation model a very precise and accurate measurements results can be achieved. In 3D measurements, however, the situation is much more sophisticated, as the influence of the Tx and Rx antennas will play a role. The influence of the antennas on localization performance is described in the next section.

6.3. Scenario related antenna selection

Antennas count among the most important passive components in every wireless system. They are responsible for matching the system impedance (e.g. $50\,\Omega$) to the free-space impedance ($\approx 377\,\Omega$) and transferring the electromagnetic signal from the electronic circuit into air. The whole process should take place with the smallest possible losses (reflections or attenuation). This task is even more complicated in the case of UWB systems, as all frequency components must be radiated or received at the same moment, and in the optimal case be influenced in the same way.

The top and side view on a typical localization scenario is presented in Fig. 6.8. Due to the diverse radiation characteristics required for different tasks, in principle there will be two to three antenna types required [19]. On the one hand, the antennas that will be mounted on the mobile units need to exhibit omnidirectional characteristics. The reason for this is that the mobile unit (e.g. a person or a robot) can rotate around its own vertical axis, but the radio-link must maintain the same quality regardless of the orientation between MU and the access point. The second kind of antennas are the directive radiators, that are implemented in the APs. Due to the fact that the access points will be in most cases placed in the corners of the room, or below the ceiling (cf. section 3.8), they will always have a defined spatial orientation and will illuminate only a selected part of the scenario. To avoid the situation in which many antennas are required to assure the full illumination of the scenario, the AP antennas should have sufficient 3 dB-beamwidth.

126

Fig. 6.8.: Top and side view on an example indoor scenario, showing the placement of the APs and an exemplary position of the MU. The favored antenna characteristics in horizontal and elevation direction are marked with circular lines.

In a typical indoor scenario, as for example an office (section 7.1) or an industrial environment (section 7.2), the AP-antennas placed in the corners, are mounted at approximately 2.5 m and 4 m in height respectively. At the same time, the height of the MU-antenna would be 1.5 m (e.g. worn by a person on a shoulder) or 2 m (e.g. on top of the fork-lift). Considering the horizontal size of those scenarios the MU-antenna will "see" the APs under the elevation angle θ ranging from $\approx 45°$ to $80°$ (office) or $30°$ to $75°$ (industrial hall) for most of the time. This information can also be incorporated while designing the antenna for the mobile user.

6.3.1. Antenna for mobile users

After the analysis of the typical scenarios, the planar wideband monopole (Fig. 6.9, left) has been developed to serve as MU antenna [16]. The radiation characteristic is close to omni-directional, with a slight dominance of the forward and backward radiation at higher frequencies (see Fig. 6.10(a)). The main radiation direction of the antenna (max. gain of 3.2 dBi) is at elevation angle $\theta=85°$ (Fig. 6.10(c)). The monopole antenna exhibits a timely compact impulse response in both the azimuth and the elevation planes (Fig. 6.10(b) and 6.10(d) respectively). This assures a minimal distortion of the radiated UWB-pulses. This is additionally almost equal for all azimuth directions ψ within an elevation range θ of $65°$ to $95°$. As this range does not completely cover the one at which the AP-antennas are expected to be "seen" in the office scenario, the system cannot profit from the maximal gain of the MU-antenna. This can become a drawback in the border zones of the scenario, where the signal is much weaker than in the central area.

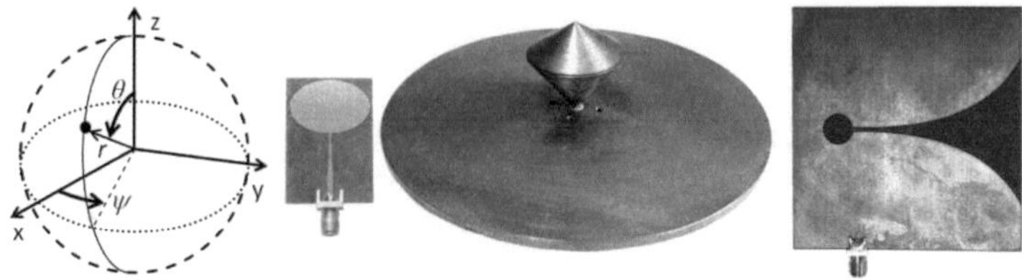

Fig. 6.9.: The the spherical coordinate system and three antennas used during measurements. From left to right: planar monopole, monocone- and Vivaldi-antenna.

Due to the described properties of the wideband monopole and the boundary conditions from the previous section, the influence of the MU antenna during localization is marginal. While using the TDOA approach, nearly the same time distortion of the AIR will be observed at all APs, and as such will be canceled due to the time-difference approach.

As mentioned before, in the industrial scenarios the APs are generally placed higher and the relative MU-AP angle becomes steeper. In this situation, the monocone antenna (Fig. 6.9, middle) is the kind of broadband radiator that is more suited for the mobile unit. Its omni-directional characteristic is more uniform in the horizontal direction, due to the rotation symmetry, compared to the one of the monopole antenna; its main beam direction is shifted up in elevation. The monocone antenna exhibits a maximal gain of 4.75 dBi in the direction of $\theta \approx 50°$. The 3 dB-beamwidth in elevation is 32° to 72° [16]. Due to these characteristics, this antenna is suitable for use during the large-scale system verification in section 7.2.

6.3.2. Access point antenna

As described at the beginning of the section 6.3, the AP-antennas, due to their placement in the scenario, should preferably illuminate the region of up to ±50° sideways the main beam direction (horizontal plane). It is desirable for the radiation to take place only to the front, so that the signals arriving from the back (e.g. through reflections from walls and ceiling) are not received. Those requirements are fulfilled by the Vivaldi antenna, depicted in Fig. 6.9 (right) [SÖ7], [7].
The Fig. 6.11(a) and 6.11(c) confirm that the desired angular range is achieved, even though the beamwidth becomes narrower towards higher frequencies. Figures 6.11(b) and 6.11(d) prove that the impulse response is very compact in both planes, preventing the broadening of the radiated UWB pulse.

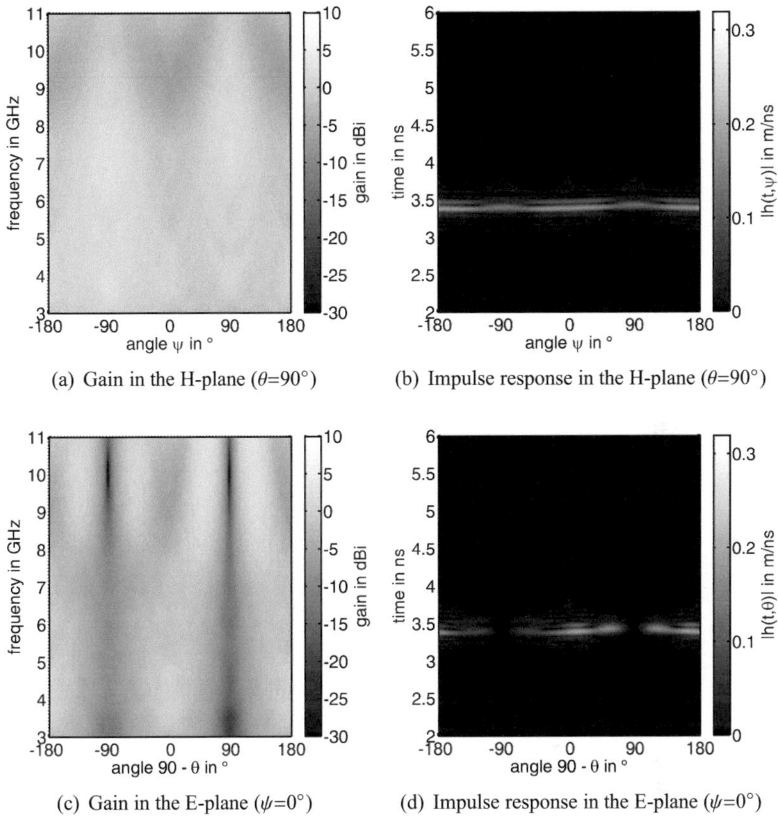

(a) Gain in the H-plane ($\theta=90°$)

(b) Impulse response in the H-plane ($\theta=90°$)

(c) Gain in the E-plane ($\psi=0°$)

(d) Impulse response in the E-plane ($\psi=0°$)

Fig. 6.10.: Characteristic of the monopole antenna: (a,c) in frequency and (b,d) time domain. The antenna was aligned in the xz-plane, pointing with the SMA-connector towards negative z-direction.

During the localization measurements, the feeding point of the antenna is taken as a reference position of the AP. However, the asymmetrical construction of the Vivaldi antenna leads to a situation where the effective signal path length, between MU and the AP reference position, depends on the direction of signal arrival. This can be observed by means of the simulated time domain characteristic presented in this section. This fact results in the uparching of the impulse response in both E- and H-planes and causes angle-dependent antenna impulse response distortion (curved shape in Fig. 6.11(b) and 6.11(d)). This effect will be discussed in greater detail in section 6.4.

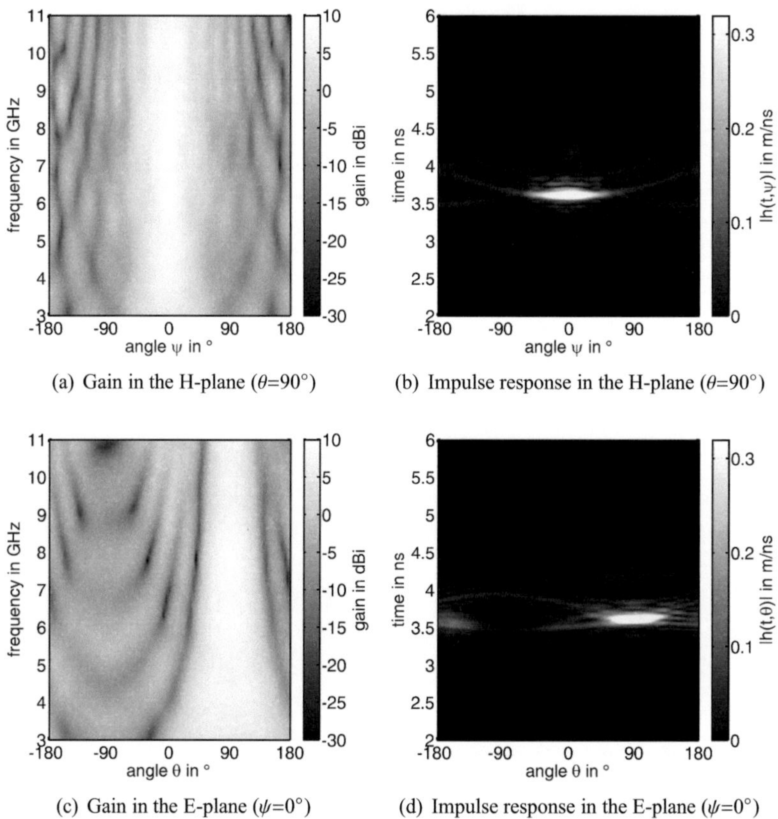

(a) Gain in the H-plane ($\theta=90°$)

(b) Impulse response in the H-plane ($\theta=90°$)

(c) Gain in the E-plane ($\psi=0°$)

(d) Impulse response in the E-plane ($\psi=0°$)

Fig. 6.11.: Characteristic of the Vivaldi antenna: (a,c) in frequency and (b,d) time domain. The antenna was aligned in the xz-plane, pointing with the tapered opening in the positive x-direction.

6.4. Influence of the antenna impulse response on time measurement accuracy

All real antennas introduce distortion during radiation [WAS09] to some extent. This deformation is angle-dependent [PZW11] along at least one coordinate. The parameter that describes the properties of the antennas in the frequency domain is the antenna transfer function $H(f,\theta,\psi)$. The corresponding parameter describing their behavior in the time domain is the antenna impulse response (AIR):

$$h(t,\theta,\psi) = \mathcal{F}^{-1}\{H(f,\theta,\psi)\}. \tag{6.4}$$

If in the frequency domain the antenna's influence on the amplitude and phase of the radiated signal changes with the coordinates, then in the time domain this will result in an angle-dependent pulse shape distortion [17]. As a consequence, the change in AIR shape and delay of its maximum will cause the time offset during the TDOA measurements to change as well. Due to this, the localization accuracy will in general be dependent on the relative angle under which the AP antenna is oriented with respect to the MU. This is discussed in more detail in this section, following the example of the Vivaldi antenna.

The effective path length that a pulse travels from the radiation source to the connector of the Rx antenna depends in general, on the reception angle (angle which the pulse enters the antenna). This happens when the feeding point does not geometrically coincide with the antenna phase center (APC). The APC is the point at which the wave physically detaches from (or couples into) the antenna. The graphical representation of this phenomenon is depicted in Fig. 6.12.

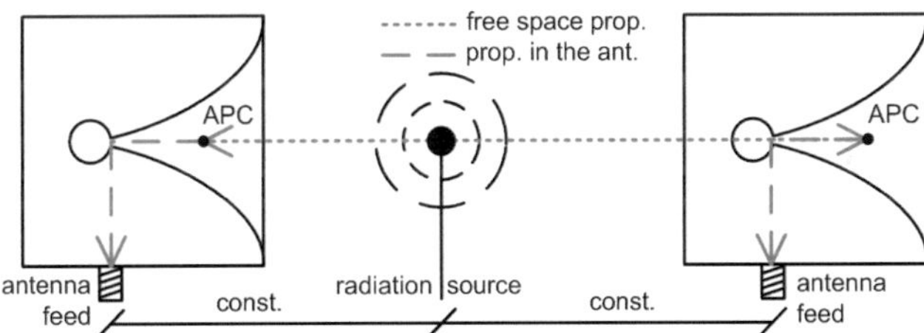

Fig. 6.12.: Propagation of the EM-wave from the source, through the antenna to the feeding point. Different path lengths of the signal during the reception from the front (left) and back (right), despite of the physically identical separation between the source and the feed.

In principle, there exist two compensation methods to minimize this effect:

I either the antenna will be exactly examined (simulated or measured) and the additional time offsets, respectively to the main beam direction, will be subtracted from the measured TDOAs depending on the arrival direction,

II or the point where the signal physically couples into the antenna (i.e. the APC) will be identified and used as the new reference position in the localization algorithm.

Both methods are described in the following two sections.

The monocone and planar monopole antennas represent special cases in this context. Although their feeding points are not identical with their APCs, these two points do not change their mutual orientation while rotating the antenna around the vertical axis. This can be verified by analyzing Fig. 6.10(b), where the AIR maximum lies constantly at 3.4 ns. In general, the APC of this kind of antennas moves slightly along the z-axis [MS08]. However, the AIR shows here no uparching (cf. Fig. 6.10(d)) in the elevation range $\theta \approx 40°$ to $140°$. Hence, the effective distance which the pulse has to travel depending on the elevation can also be assumed to be nearly constant here. As a consequence, this effect can be safely neglected for the omnidirectional antennas from section 6.3.1.

6.4.1. Look-up table compensation method

The basic requirement for the application of the method I is the knowledge about the three-dimensional, angle-dependent delay of the AIR's maximum of the used antenna. To achieve this, a set-up similar to the one in Fig. 6.12 is used. The antenna under test is rotated in two dimensions around the feeding point and for every angle configuration the received pulse is recorded. The delay of the impulse response is determined relatively to the main beam direction. In Fig. 6.13, the angle-dependent delay pattern of the Vivaldi antenna is shown, which results from an evaluation of the data acquired in the above described way. The antenna main lobe points, in this situation, in the direction $0°$ elevation and $90°$ azimuth ($\theta = 90°$ and $\psi = 0°$, compared to Fig. 6.11).
The information about the AIR delay pattern can be obtained either by means of a simulation or by measurements. This aspect is considered at the end of this section.

The information contained in Fig. 6.13 can be stored in the form of a look-up-table (LUT). In order to eliminate the angle-dependent delay of the antenna impulse response, with the help of information stored in the LUT, an iterative approach can be used [17]. The flow diagram of this algorithm is depicted in Fig. 6.14.

The requirement for this algorithm is the knowledge about the spatial orientation of the AP antennas. This can easily be assured during the system

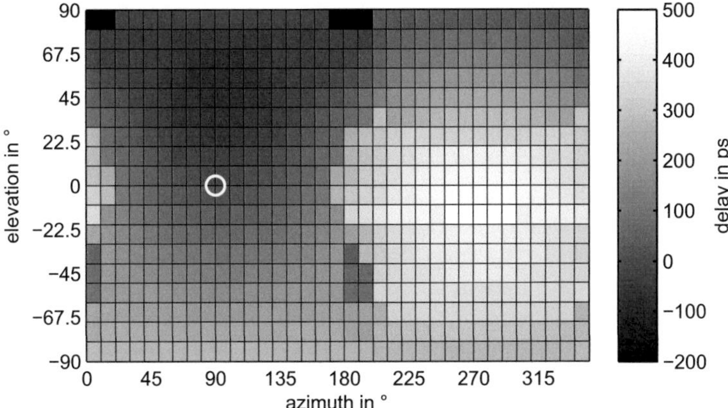

Fig. 6.13.: Angle-dependent AIR delay pattern of the Vivaldi antenna, resulting from simulation. The delays have been determined using the maximum of the PG5 pulse, after conversion to baseband. The main beam direction is marked with the white circle (az. $= 90°$ and el. $= 0°$).

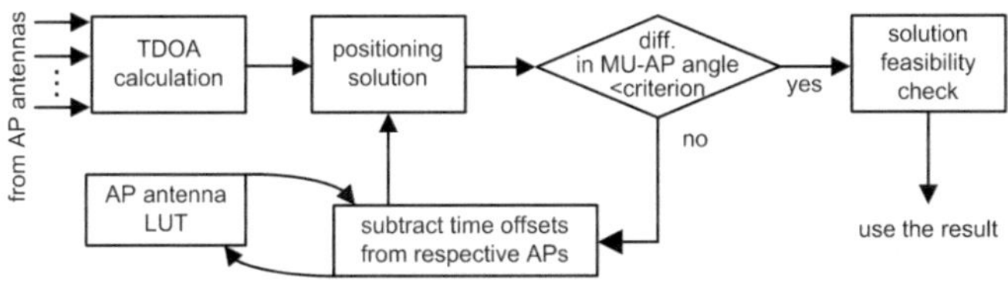

Fig. 6.14.: Flow diagram of the AIR compensation method, based on LUT.

deployment phase. In a localization system with multiple access points, the impulse radiated by the MU is received by the APs and the time differences are calculated. After obtaining a valid TDOA measurement, the first step in the algorithm is the standard calculation of the MU position. This leads to a solution which serves as a starting point for the iteration. Now, knowing the approximate MU location, the relative angles between the reference direction of every AP and the estimated MU position are calculated. For those angles, the time correction factors for each AP are obtained from the look-up-table. After subtracting the correction factors from the original time differences, the new position is calculated. This operation is re-iterated until a break criterion is met, i.e. when the change in the relative angles in two consecutive iterations is smaller than a certain value. Another criterion could be the change in the calculated positioning solution [18].

Experimental verification of the LUT method

For the purpose of comparison, the model of the Vivaldi antenna [SÖ7], created in CST Microwave Studio simulation software [Com12], was used. In the simulation, the feeding point (connector) of the antenna is placed in the origin of the coordinate system and the ideal signal (E-field) probes are placed in the far field, on an outline of the circle with a constant distance to the origin. Starting from the main radiation direction, the probes were placed in 10° separations in the H-plane. At each probe, the impulse response of the antenna was recorded, where the UWB pulse (identical with the one used during measurements) has been applied as an excitation.

The measurement system, used for the verification of the simulation results, consisted of a UWB transmitter, two antennas (monocone as a transmitting and Vivaldi as receiving one), the UWB autocorrelation receiver and an oscilloscope. The schematic representation of this setup is shown in Fig. 6.15(a). The impulse generator, triggered from the oscilloscope, transmits the signals via the monocone antenna to the Vivaldi antenna placed on a rotary tripod in 1 m distance. The received impulse is processed by an autocorrelation receiver and recorded with the oscilloscope (Agilent DSO-81204A). This measurement is performed at the same angles as in the simulation.

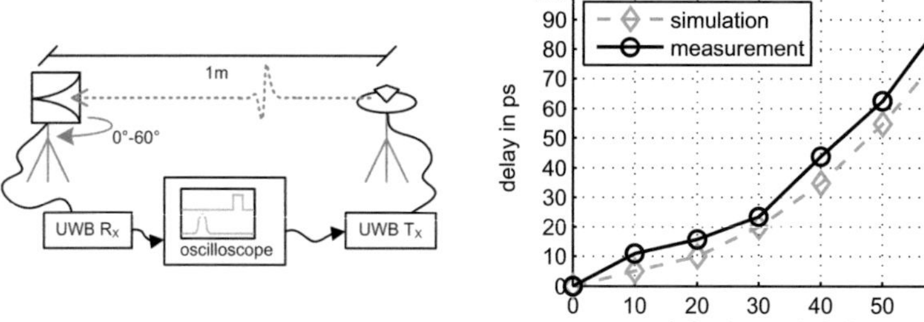

(a) LUT verification - measurement setup

(b) comparison between measurement and simulation

Fig. 6.15.: Comparison of the measured and simulated AIR peak delay of the Vivaldi antenna as a function of the angle in the H-plane.

Since the UWB receiver is realized as an auto-correlation receiver, the oscilloscope records already the baseband signal after the correlator. In order to be able to compare the simulation results with the measured ones, the simulated impulse response is processed in the same way in Matlab: after squaring and low-pass filtering the baseband envelope is obtained.

The other difference between the measurement setup and the simulation is the use of the monocone antenna instead of an ideal E-field probe. This however, does not have any influence on the measurement results, since the relative and not the absolute time shift of the received AIR maximum is of interest. During verification, the deflection angles up to $60°$ were investigated, as above this limit the amplitude of the received signal is too weak to clearly identify the maximum.

In Fig. 6.15(b) the simulation results are compared with the ones obtained by measurement. The deflection angle is defined as the angle between the main lobe direction and the measurement angle. It can be observed that the principle shape of the measured and the simulated curves fits very well, though in the measurements a larger delay is obtained. However, the value of this offset is nearly constant and does not change with increasing deflection angle. The average delay difference is smaller than 7.2 ps. The full description of the simulation and measurement procedure can be found in [Sch11] and [17].

This verification proves that the simulation can safely be used for the creation of the 3D look-up table of the antenna impulse response.

6.4.2. Compensation method based on phase center calculation

A method that is more elegant and less circuitous than the one presented in section 6.4.1, is to determine the actual phase center of the antenna (either by simulation or measurement) and to use it directly as the reference position for the localization system. The unique identification of the phase center coordinates is however possible only in case of some antennas. The APC of e.g. the Vivaldi antenna is not stable over frequency and moves along the symmetry line of the tapered opening (same as main lobe direction) [WJB$^+$06]. This happens because the radiation condition for different frequencies is fulfilled at physically different points. It is thus not possible to calculate a unique point where the impulse is radiated, as UWB signals consist of a very broad range of frequencies. The ratio of wavelengths between the lowest and the highest frequency components is in case of the FCC regulation as big as 3. This results in phase center position variations larger than 20 mm [EZL$^+$07].

To better assess the feasibility of this correction method, the phase center of the Vivaldi antenna from section 6.3.2 has been calculated using CST Microwave Studio. This was done for f_M=6.85 GHz, which is the center frequency of the FCC spectrum. According to the results, the reference point would have to be transformed by a vector $(x,y,z) = (2.1, 25.3, 41.0)$ mm, with respect to the feeding point (main beam in +y-direction). In the experiment described in [MZM$^+$08] (measurement of the Vivaldi antenna in anechoic chamber), it was shown that between 6 GHz to 10 GHz, the phase center changes its position by ≈ 3 mm, hence in the full FCC band even larger fluctuations are expected. Furthermore in [MZM$^+$08], the variation as a function of angle was reported to reach 1 cm at $\pm 40°$, however, no compensation algorithms were proposed.

In section 7.1, the effectiveness of the look-up table and phase center compensation methods is compared using real measurement data.

6.5. Localization error bound

The Cramer-Rao lower bound (CRLB) is one of the metrics often used to estimate the best case performance in terms of precision in positioning systems. In general, the CRLB describes the lowest achievable limit of the variance of an estimator and is independent on a particular estimation algorithm. This method can be directly adopted to the UWB localization system, where the precision of the calculated MU position shall be related to the TDOA measurement statistic. As also indicated in [PAK$^+$05], the CRLB is the function of:

- the number of known reference nodes (access points) and type of localization: 2D or 3D,

- the geometrical orientation of the APs and the position of the MU,

- the interconnection of the APs (e.g. whether the redundant measurements are used or not),

- the statistical distribution of the measurement values (here: time differences).

In this section, the lower limit on the position estimation variance in a TDOA localization system is derived using the CRLB. The extension to a 3D localization case is based on the approach presented in [XML08], where only the 2D situation was considered. In the derivation, the assumptions and notation described below have been used.

In general, the notation and indexing from chapter 3 is used. To reduce the number of used indexes, the 3D positions of two arbitrary APs in the Cartesian coordinate system will be denoted shortly as r_i and r_j instead of $\vec{r}_{E_i}$ and $\vec{r}_{E_j}$, as in equation (3.2). The total number of access points used for localization is N, and only the non-redundant TDOAs are considered (e.g. differences calculated only to the 1st AP, $i = 1$). The true position of the MU is r_S (instead $\vec{r}_S$), whereas its estimated position is $\hat{r}_S$, instead of the complete notation $\hat{\vec{r}}_S$. The true distance difference between the 1st and j^{th} AP and the MU is described by equation (3.7). The measured distance $\Delta \tilde{d}_{ij}$ is however afflicted by the noise w of each involved AP, according to (3.12). All the measured TDOAs (or distance differences) can be summarized in the vector: $\tilde{\vec{\rho}} = [\Delta \tilde{d}_{12}, \Delta \tilde{d}_{13}, ..., \Delta \tilde{d}_{1N}]^T$ ($\tilde{\rho}$ in simplified notation).

As was shown in section 6.2, the time measurement error $w_{i/j}$ of the TDC unit is normally distributed. Assuming for simplification that $w_{i/j}$ is zero-mean (despite the findings from section 6.2.1) it can be written as $\mathcal{N}(0, \sigma_{i/j}^2)$. The difference of two independent, normally-distributed variables (Δw) is also normally distributed: $\mathcal{N}\left(0, (\sigma_i^2 + \sigma_j^2)\right) = \mathcal{N}\left(0, \sigma_{\Delta w}^2\right)$.

The probability of obtaining a certain value of $\Delta \tilde{d}_{ij}$ by means of a measurement is:

$$p(\Delta \tilde{d}_{ij}) = \frac{1}{\sqrt{2\pi\sigma_{\Delta w}^2}} \exp\left(-\frac{(\Delta \tilde{d}_{ij} - \Delta d_{ij})^2}{2\sigma_{\Delta w}^2}\right). \tag{6.5}$$

Knowing that time differences are calculated with respect to one reference receiver with index $i = 1$, hence only non-redundant measurements are utilized (in contrast to [XML06]), the conditional joint PDF is:

$$p(\Delta \tilde{d}_{ij} | x_S, y_S, z_S) = \prod_{i=1; j=2}^{N} p(\Delta \tilde{d}_{ij}) = p(\tilde{\rho} | r_S), \tag{6.6}$$

where $\tilde{\rho}$ is the previously introduced vector of TDOA pseudo-ranges. Using this, the Fisher Information Matrix (FIM) is defined as:

$$I(r_S) = I(x_S, y_S, z_S) = E\left\{\left(\frac{\partial \ln(p(\tilde{\rho}|r_S))}{\partial r_S}\right)\left(\frac{\partial \ln(p(\tilde{\rho}|r_S))}{\partial r_S}\right)^T\right\} \tag{6.7}$$

$$I(r_S) = \begin{bmatrix} I_{11} & I_{12} & I_{13} \\ I_{21} & I_{22} & I_{23} \\ I_{31} & I_{32} & I_{33} \end{bmatrix} = \begin{bmatrix} I_{xx} & I_{xy} & I_{xz} \\ I_{yx} & I_{yy} & I_{yz} \\ I_{zx} & I_{zy} & I_{zz} \end{bmatrix}. \tag{6.8}$$

The Cramer-Rao lower bound states, that the variance of the unbiased position estimation of the MU cannot be smaller than the inverse of the FIM:

$$C_{\hat{r}_S} \geq I^{-1}(r_S) \quad \Rightarrow \quad C_{\hat{r}_S} - I^{-1}(r_S) \geq 0, \tag{6.9}$$

where the covariance matrix of the position estimation of the MU is defined as:

$$C_{\hat{r}_S} = \mathrm{cov}\{\hat{r}_S\} = E\left\{(\hat{r}_S - r_S)(\hat{r}_S - r_S)^{\mathrm{T}}\right\}. \tag{6.10}$$

The position error bound, according to Cramer-Rao satisfies this relation:

$$\sigma^2_{\mathrm{MU}}(\tilde{r}_S) = \mathrm{var}(\tilde{x}_S) + \mathrm{var}(\tilde{y}_S) + \mathrm{var}(\tilde{z}_S) \geq \mathrm{trace}\left(I^{-1}(r_S)\right). \tag{6.11}$$

According to the calculations presented in the appendix A, the general expression for an arbitrary element of the Fisher Information Matrix from (6.7) can be written as:

$$I_{kl} = \frac{1}{\sigma^2_{\Delta\omega}} \cdot \sum_{i=1;j=2}^{N} \left(D_{k_i} - D_{k_j}\right)\left(D_{l_i} - D_{l_j}\right), \tag{6.12}$$

where $k, l \in \{x, y, z\}$. The sub-function D, e.g. for coordinate x and i^{th} AP, has the following form:

$$D_{x_i} = \frac{(x_S - x_i)}{\sqrt{(x_S - x_i)^2 + (y_S - y_i)^2 + (z_S - z_i)^2}} = \frac{(x_S - x_i)}{\|r_S - r_i\|}, \tag{6.13}$$

hence is identical with (3.27) and can be interpreted as a direction vector.

Using the closed form expression for the elements of the FIM from (6.12), it is now possible to calculate its inverse (cf. appendix A). The CRLB can be then calculated as the trace of the FIM inverse according equation (6.11).

Concluding remarks on Cramer-Rao bound and its derivation
As indicated during the derivation, the assumption about unbiased TDOA measurements has been made, what is only partially correct. According to results from section 6.2.1, the trigger offset introduces a negative bias on time measurements. This implies that the mean value of $p(\Delta\tilde{d})$ should be modeled as linearly distance-dependent. The procedure for calculation of the CRLB provided the knowledge of the measurement bias function is given e.g. in [MH06]. In this study however, a correction method has been proposed that compensated the trigger offset effect (see section 6.2.1). In practice, should

the trigger offset compensation be imprecise, this will result in too optimistic CRLB.

Furthermore, in equation (6.5) a constant TDOA variance is used, which contrasts with the measurements results from section 6.2.2. On the other hand, the time-difference variance changes only slightly in the range up to 10 m, within which the tests from chapter 7 have been performed. Considering all the above, the here derived CRLB for a 3D TDOA localization system is a simplified model that offers a benchmark to assess the precision of the conducted positioning experiments. The CRLB cannot however give any information regarding the localization accuracy, as unbiased measurements were assumed during derivation. The comparison results confirming the very good agreement of the above theory with the measurements are presented in sections 7.1.1 and 7.2.2.

7. Verification of the designed IR-UWB localization system

In the previous chapters, the single components of the impulse-based UWB localization system have been introduced and thoroughly investigated with regard to their performance. Alongside with them, the position estimation algorithms, prediction and correction methods have been developed and examined. All those subsystems have then been merged to create a complete indoor localization system. In this chapter, the results of the verification measurements performed with this experimental system are presented and analyzed. This includes the tests conducted in a laboratory environment, representing a small-scale application. These are described in section 7.1. The main focus in this part is to test the accuracy correction methods and the impact of the time-averaging on the localization precision. This is followed by the experiment performed with the upscaled system in a quasi-industrial environment in section 7.2. Finally, in section 7.3, a feasibility study on the fusion of a UWB localization system and an Inertial Measurement Unit based navigation system is presented and backed-up with measurements.

7.1. Verification of the system in a small-scale indoor scenario

The localization system described in this section is based on the architecture presented in section 6.1. The system configuration was the following:

- the mobile user transmits pulses of 1 V peak-to-peak amplitude at a rate of 1 MHz and is equipped with a planar monopole antenna.

- The access points are the receivers (ACR) fitted with Vivaldi antennas. They are connected to the time measurement unit (TDC) using cables of equal lengths, making the calibration procedure significantly easier (see equation (6.1)).

- The TDOA readings are collected from the TDC by the PC and used for calculating the 3D positioning solution for the MU.

This setup was deployed in a laboratory room[12]. The model of the scenario, alongside with the photo of the measurement setup are depicted in Fig. 7.1. In the left picture, the positions of all five APs are marked with circles; the exemplary one of the MU with a white rhombus in the middle. In the right photograph, the part of the measurement setup is visible: MU and one of the AP antennas on tripods (right) and trigger generator together with TDC module (left). During the measurements, the lab equipment as well as the furniture were left within the scenario on purpose to ensure realistic testing conditions.

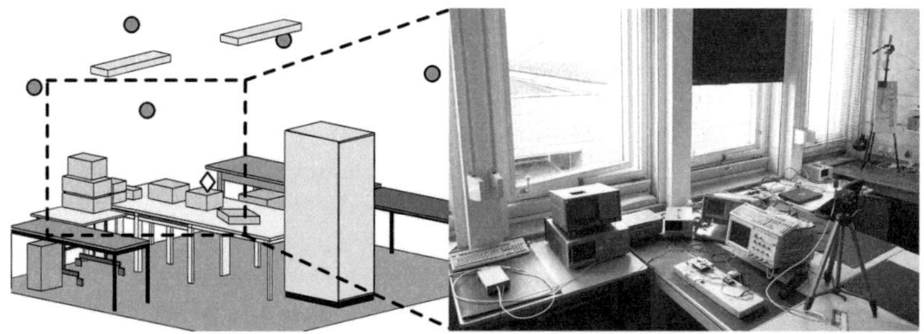

Fig. 7.1.: Model (left) and the photograph of the laboratory room (right).

The APs were placed on the outline of the $4\,\text{m} \times 6\,\text{m}$ large area. Their heights were chosen in a way to achieve a most uniform distribution of DOP-values, shown in Fig. 7.2. The AP antennas 1 and 2 were pointing along the x-axis in the positive, and 3 together with 4, in the negative direction. The 5th antenna was pointing along the y-axis (positive direction).

The mobile user was placed at four different positions on the table (marked P-1 to P-4), in the middle part of the room (diamond markers). The reference positions of the APs, as well as those of the MU, were determined by means of the 3D distance measurements relative to the room walls and the floor. The nominal accuracy of the laser distance measurement device, used for this purpose, was 1 mm. It is, however, likely that the overall accuracy of the reference positions is lower due to human error.

With the calibrated system in place, 1000 time difference measurements at each MU position were performed. They were repeated for different amount

---

[12] Institut für Hochfrequenztechnik und Elektronik at the KIT-campus, building 30.34

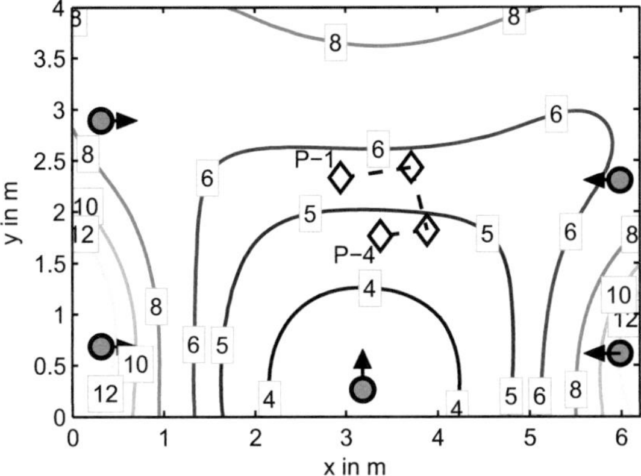

Fig. 7.2.: Distribution of the PDOP values (top view) in the lab. Dots represent the AP antenna positions and the arrows the alignment of their main beam direction. The rhombi are the investigated MU positions.

of time difference averaging at the TDC. In the following two sections, the aspects of the system accuracy and precision are addressed.

7.1.1. Influence of time averaging on localization precision

In this section, the aspect of localization precision dependency on the averaging during the time difference measurements is analyzed. The precision describes the repeatability of the results. In this case, it refers to the spread of the positions, calculated from multiple TDOA measurements, performed in static condition (unchanged MU and APs positions).
In the magnified part of Fig. 7.4, the scattering of the positioning solution is clearly visible. This is caused by the uncorrelated noise at the TDC during the TDOA measurements. In addition, another prominent effect can be observed: the scattering is much stronger in the vertical direction than in the horizontal one. This is directly influenced by the geometry of the access points relative to the mobile user.

The rate of the scattering or precision can be estimated using the Cramer-Rao lower bound, described in section 6.5. This allows the determination of the minimum achievable variance of the solution, assuming unbiased TDOA measurements with a known PDF. The relation of the achieved localization precision and the CRLB has been investigated in dependency of TDOA averaging. The results are depicted in Fig. 7.3.

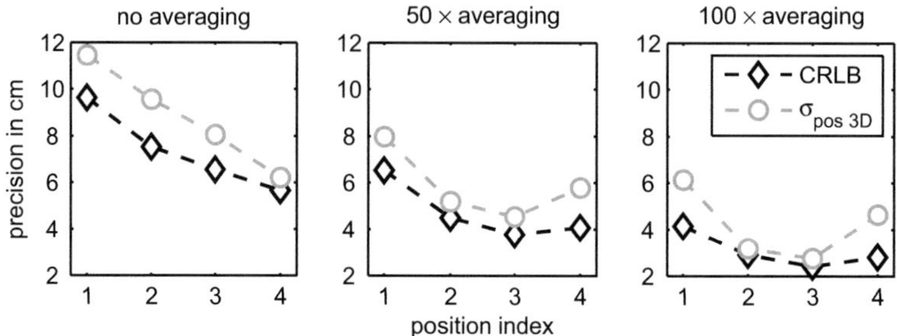

Fig. 7.3.: Comparison of the 3D localization precision at four investigated MU positions with the Cramer-Rao lower bound and the dependency on the TDOA averaging.

For each point on these three curves, determined by measurement (circle marker), 1000 TDOA values have been used. The time difference measurements were taken: without averaging (left), with 50- (middle) and 100-fold averaging (right). No additional corrections were applied. First of all, it is evident that the standard deviation $\sigma_{pos\,3D}$ of the positioning solution never crosses the theoretical lower limit indicated by the CRLB. As the number of averages per TDOA measurement increases, the variation of the sample values reduces, causing both the CRLB and $\sigma_{pos\,3D}$ to decrease. This change depends however also on the system geometry, what explains the different rate of precision improvement at the four investigated locations.

It has to be stressed that the improvement in terms of the lower level of variations does not necessarily imply an improvement of the accuracy, e.g. if the measured TDOAs were biased.

7.1.2. Validation of the accuracy and precision correction methods

In this section, the localization results obtained during the small-scale localization experiment are analyzed in terms of the positioning accuracy and precision. The correction methods, listed mainly in chapters 3 and 6, leading to improvement of these two parameters are evaluated.

The results are summarized in table 7.1. The point of departure for the consideration is the evaluation of the AP constellation quality at each MU position.

The DOP values describing this property are listed in the second column. Knowing the standard deviation of the TDOA measurements at these positions (no averaging), the precision prediction can be performed according to (3.65) (column 3). According to [Nee09], this kind of prediction can be considered as an effective approximation of the RMS accuracy (combined accuracy and precision), when a biased estimator is suspected. This prediction coincides very well with the mean positioning error obtained from 1000 un-averaged measurements in column 4. In contrast to CRLB, this prediction is not a hard bound.

Table 7.1.: Positioning statistics for the scenario from Fig. 7.4 (values in cm).

Position	corresp. DOP (no unit)	predicted precision	calculated mean error	mean error 100-avg.	removed outliers	LUT	phase center
1	5.55	11.97	10.15	6.29	3.52	2.83	2.41
2	5.73	9.15	9.27	7.94	7.89	3.36	4.76
3	4.74	8.22	7.38	3.01	1.80	1.35	1.53
4	4.59	8.84	8.17	5.27	3.26	3.19	3.43
ø	-	9.55	8.74	5.63	4.11	2.68	3.03

Due to the fact that the TDOA measurement error of the TDC is normally distributed (cf. section 6.2), the localization precision can be improved by using averaging (see previous section). In column 5, the average positioning error obtained from 1000 calculated positions is presented. In turn, for calculation of each of these positions 100-fold averaged TDOA measurement has been used.

The *velocity filtering* (cf. section 3.7), that performs the plausibility check in terms of the MU speed, based on the distance difference between two adjacent positions and the elapsed time, leads to further improvements (column 6 of table 7.1). The rate of improvement achieved by the TDOA averaging and the velocity filtering is not equal for each MU position. This is caused by the fact that as the precision is successively improved the limited accuracy (e.g. small errors in the measured AP and MU reference coordinates) begins to dominate the overall positioning error.

The last step is the compensation of the angle-dependent pulse delay, introduced by the AP antenna. The application of the iterative correction algorithm, proposed in section 6.4.1, leads to the results listed in column 7. The last column in table 7.1 contains the results of the AIR correction according to the phase center method. The second one leads to a slightly lower accuracy compared to the LUT method, due to the issues described in section 6.4.2. However, the correction according to phase center method is less time-consuming, as no iterations are required. Both methods allow the final mean

3D error to be achieved at a remarkably low level of only $\approx 3\,\mathrm{cm}$. The last two steps correct the time bias caused by the antenna, hence improving the accuracy. The amount of accuracy improvement offered by them is however dependent of the initial orientation of the AP antennas with regard to the mobile user.

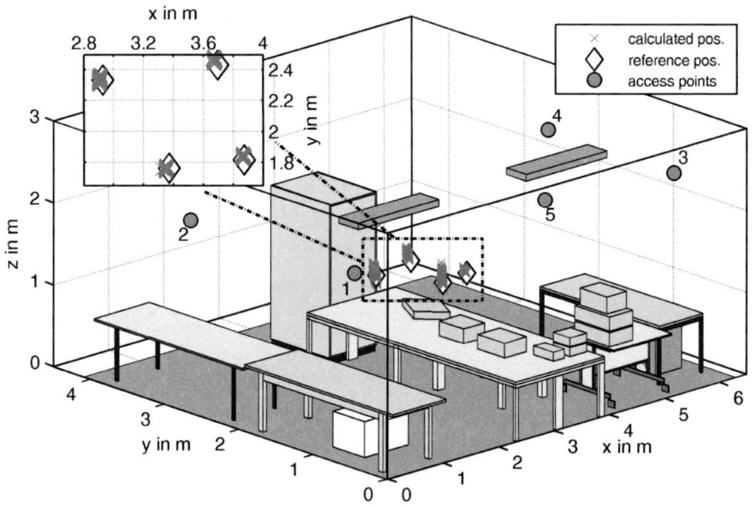

Fig. 7.4.: Visualization of the localization results (crosses), achieved during the laboratory-scale experiment.

The localization results, plotted into the scenario model, are presented in Fig. 7.4. For this purpose the corrected data has been used. The solutions (marked with crosses) lie very close to their reference positions (white diamond shaped markers). The MU was placed close to the center of the AP-constellation and the single MU-AP distances were similar, hence the effect of the trigger offset, described in section 6.2.1, does not play a role here. Nevertheless, the average localization solutions are still slightly biased, which can be due to the inaccuracies during the reference position measurements.

7.2. Verification of the system in a large scale indoor scenario

After successful validation of the UWB localization system functionality and proving the effectiveness of previously proposed accuracy and precision enhancement methods in a laboratory-scale, the upscaling potential of the system shall be investigated. To achieve this, the system has been modified and installed in an general-purpose assembly hall[13]. The system architecture has been slightly altered to enable operation in this environment:

- to provide the better signal coverage, the transmitted pulse amplitude of the UWB pulses has been increased to 4 V peak-to-peak,

- an additional access point has been used to enhance the distribution of the DOP values,

- in contrast to the previous experiment, the APs acted as Tx and MU received the pulses,

- the AP synchronization network was adjusted to cope with much larger signal delays, compared to the lab scenario.

The rectangular trigger signal, generated at a PRF of 1 MHz, was first split into 6 branches by a custom-built-hub; those were then connected to each AP. The connecting cable lengths were different (ranging from 8 m to 50 m), hence the APs were triggered sequentially. The selection of the triggering PRF allowed for the acquisition of up to 10 000, 100-fold averaged, TDOAs every second. This update rate offers a great potential and would easily fulfill the requirements usually put on real-time localization systems (RTLS). During the measurements the MU was mounted on a trolley and was moved between the measurement positions maintaining a 1 m height of the antenna (monocone) above the ground.

The model of the measurement scenario is presented in Fig. 7.5. The positions of the APs are marked with dots, numbered from 1 to 6 according to the triggering order. The coordinates where the MU was placed during measurements are marked with rhombi. The test area has a size of 9.5 m × 10 m and is placed in the central part of the building.

The photograph of the system deployed in the wbk-hall is presented in Fig. 7.6: The access points (top and bottom left), the test area (right) and the mobile user (trolley, visible in all three pictures). In order to obtain the

---

[13] wbk Institute of Production Science at the KIT-campus, building 50.36

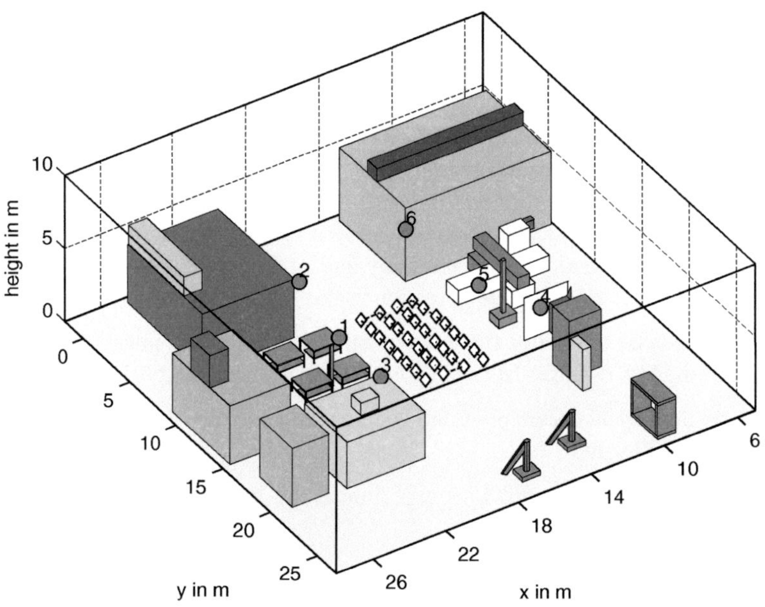

Fig. 7.5.: The model of the wbk-hall - isometric view. The positions of the access points are marked with dots and numbered according to the triggering order. The investigated positions of the mobile user are represented by the white rhombi.

exact coordinates of the MU that will serve as a reference during the accuracy analysis, the *total station* has been used. The total station is an electronic theodolite (instrument for precise angle measurements in the horizontal and vertical planes), integrated with a distance meter. This allows the measurement of the slope distances (spherical coordinate system) from the instrument to a particular point. The total station used during the experiment achieves a 3D accuracy better than 1 mm. It is visible in the top left part of Fig. 7.6.

The distribution of the APs in the scenario was done considering the remarks from section 3.8, with the goal of achieving a uniform as possible HDOP distribution. The optimization of the AP constellation with regard to VDOP was not directly possible due to the construction limitations in the building. The resulting distribution of PDOP in the scenario is depicted in Fig. 7.7. The values of DOP span in the relevant area from 3 to 9.

148

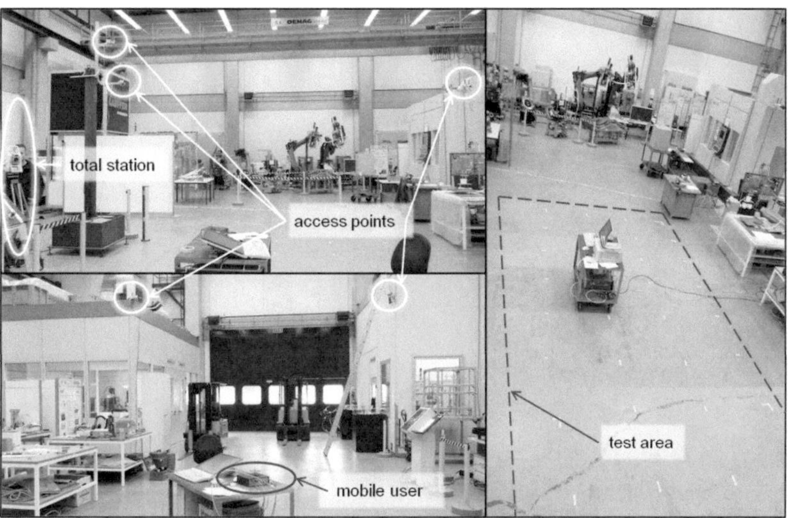

Fig. 7.6.: Photograph of the measurement setup deployed in the wbk-hall.

At each position in the scenario, 1000 TDOA measurements were recorded, where each measurement is a result of 100-fold averaging performed at the TDC. The analysis of the results is separated in two parts and conducted in following sections. In the presented results the two correction methods have already been applied on the raw data: antenna phase center and the trigger offset (based on path loss, including the MU and AP antenna gains).

7.2.1. Influence of time- and position-averaging on accuracy

In this section, a closer look is taken not only at the achieved overall accuracy, but also at its horizontal and vertical components. Furthermore, the aspect of determining the accuracy with two different types of averaging is treated.

The accuracy, calculated based on the average positioning solution, is presented in Fig. 7.8 and marked with circles. In addition to the 3D accuracy (bottom plot), its horizontal (xy-plane) and vertical components have been calculated. The diamond markers indicate the 3σ-value of the expected localization error (no hard bound), based on the analysis of DOP values and the TDOA measurements statistics [BH99, Nee09]. The calculated positions clearly correlate with the predictions.

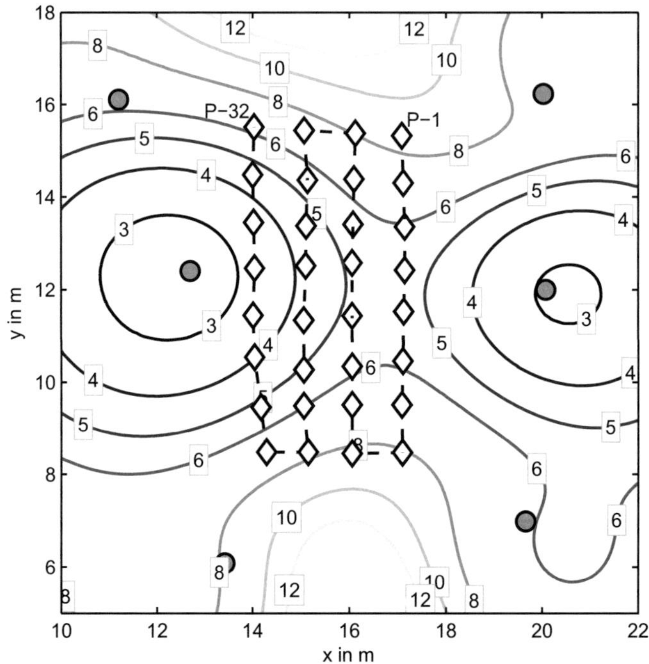

Fig. 7.7.: Top view on the scenario with the distribution of the PDOP values, represented as iso-lines with labeled values. The white rhombi represent the mobile
user reference positions, starting at P-1 is (top-right) and ending at P-32
(top-left).

The achieved average accuracy (over 32 MU positions) of the 2D localization
in this experiment was 3.44 cm. It is thus significantly better than the one of
the 3D solution, exhibiting an average accuracy of 11.37 cm. The mean vertical accuracy is 10.4 cm. The disproportion between horizontal and vertical
accuracy is however well explicable with the analysis conducted in section
3.8. During the experiment, all 6 APs were placed in 3 m to 3.5 m height. In
order to achieve a better height accuracy, the UWB infrastructure should be
extended by adding at least one additional AP, placed centrally above all other
ones (analogous to Fig. 3.8(c)).

A very important question from a position-estimation time-efficiency point of
view is how to calculate the average accuracy from the measured TDOAs. Ba-
sically, there are two ways to do this: either to estimate the position from each
set of time data and to calculate the average position error (as done above),

150

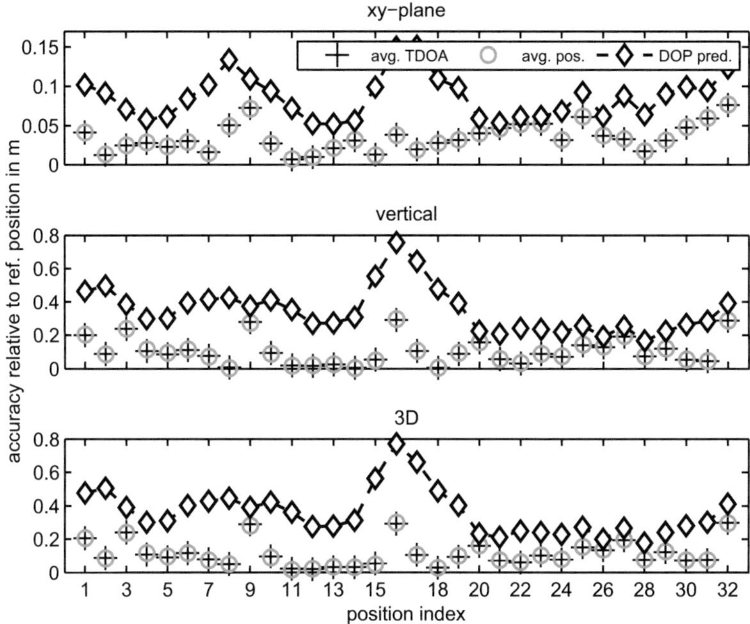

Fig. 7.8.: Comparison of the estimated (black diamond markers) and achieved accuracy. The actual achieved accuracy is marked with pluses when the average TDOAs were used, and circles for the average solution resulting from 1000 calculations.

or to first average the TDOA data and from this estimate only one position. Both methods are compared in Fig. 7.8, where the results of the second listed method are marked with plus signs. It is evident that the results are nearly the same and the calculated average difference between both solutions is as small as 3 mm. This result is to be expected, provided that the position estimator itself is unbiased. In both cases 100 000 TDOA measurements were used to achieve the final result. The second method is much more time-efficient, as the iterative position estimation algorithms have to be used only once. With 8 ms time required for single position calculation (cf. section 7.2.3), both approaches are capable of real-time operation. The time saving offered by the second approach is in linear proportion to the number of acquired TDOA data sets.

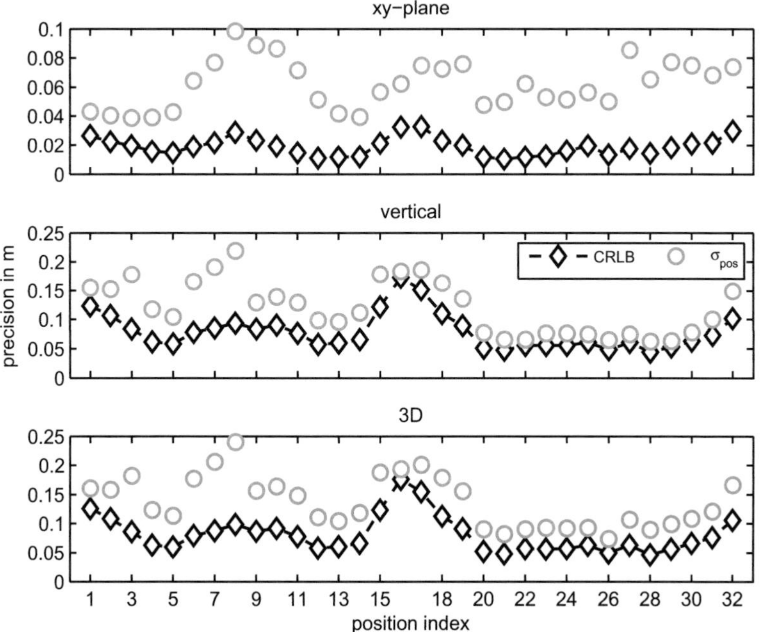

Fig. 7.9.: Comparison of the predicted and achieved precision. The predicted lower
limit of the precision - the CRLB - is plotted with black diamond markers.
The actual achieved precision for 1000 calculations is marked with circles.

7.2.2. Analysis of the horizontal, vertical and 3D precision

The position standard deviation has been calculated for each position on the
MU route and depicted in Fig. 7.9. The σ_{pos} has been split into its horizontal
(top) and vertical components (middle). As predicted, due to the AP constellation, the achieved average horizontal precision (σ_H) with 6.2 cm is much
better than the vertical one (σ_V) with 12 cm. The σ_V follows the calculated
CRLB very closely, meaning that there is not much room for improvement in
this particular AP setup. The σ_H could be most probably improved by further
averaging, however the amount of measurement data is not sufficient to prove
this hypothesis. The overall precision σ_{3D} is 13.7 cm and differs from the
theoretical limit given by the CRLB by 5.5 cm in average.

This experiment conducted in a quasi-industrial environment has proven the scalability of the UWB localization system. This can be done without the significant loss of accuracy, particularly the horizontal one. The 3D accuracy can be improved by adjusting the geometrical configuration of the APs. From the practical view point, the horizontal accuracy and precision are much more important in real applications.

7.2.3. Performance of the range-comparison RAIM in conjunction with Bancroft algorithm

The RAIM methods, used to detect time measurement offsets present at access points, have been introduced in section 3.6. The range-comparison approach has been also identified as the most suitable one for application during indoor localization. The functionality of the RCM RAIM, when using simulated data, has been validated in [3]. In this section, this is demonstrated using the measurement data.

During the creation of a decision matrix, based on which the biased AP will be identified, multiple four- and five-combinations of APs are used to calculate positioning solutions. As indicated in the example from table 3.2, in the case of 6 APs 45 position calculations are performed before the decision regarding the integrity of measurements is made. Due to this fact, the calculation of the MU position using non-iterative algorithms can significantly reduce the calculation times.

For the investigation of the RAIM algorithm the MU position P-13 from the above scenario has been chosen and the time bias has been successively added to the AP 6 in 100 ps steps. For 1000 TDOA values measured at P-13 and for all offset values the positioning solutions have been calculated. The results depicted in Fig. 7.10 show the mean position error (left) and the average calculation time[14] per position (right), as a function of bias. The used detection thresholds for classification of position residues were 15 cm and 30 cm.
The calculation of the final position has been carried out for three different cases:

1. the RAIM remained inactive: the initial solution was done using Bancroft algorithm, followed by the iterative calculation (circle markers).

2. RAIM was activated: for the position calculations within RAIM only the Bancroft algorithm was used. After identification of the biased AP

---

[14]calculations performed on a desktop PC with 2.83 GHz Q9550 processor and 8 GB RAM

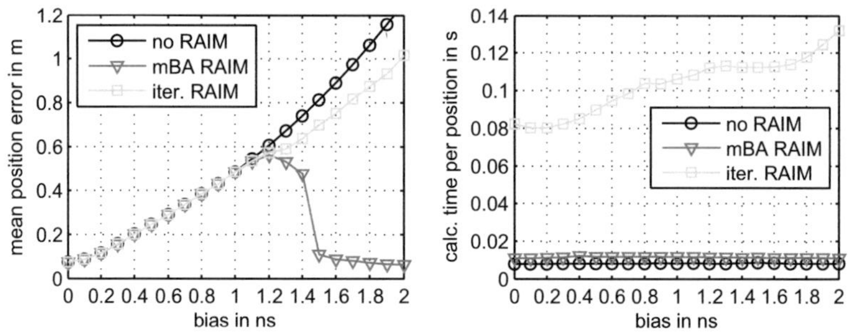

(a) Influence of RAIM on calculated error   (b) Advantage calculation time per position

Fig. 7.10.: The bias detection and calculation time performance of the implemented
RAIM algorithm in conjunction with Bancroft algorithm.

the final calculation was done using both non- and iterative method
(triangle markers).

3. RAIM was activated and all position calculations (within the RAIM and
the final one) have been done using both methods (square markers).

The results in Fig. 7.10(a) indicate, that without using RAIM the position
error grows with increasing time bias as suspected. The activated RAIM,
using only non-iterative solution, is sensitive to offset and identifies the biased
AP after the upper threshold is crossed. The error curve does not however
drop abruptly at this offset value. This is caused by the TDOA noise in the
measurements, that leads to a situation in which first at a larger bias the
calculated residues cross the threshold and the AP is identified as not useful.
Calculating the mean over 1000 samples results in this kind gradual decrease.
In the third approach only a small improvement is observable. Additionally,
the use of an iterative algorithm during the RAIM calculations causes often
the solutions to drift too far apart, what makes the clear identification of the
biased AP not possible.

The Fig. 7.10(b) shows the average time required to calculate one final MU
position, depending on the used method and bias value. The shortest computation time of 8 ms is reached without using the RAIM. Employing only the
Bancroft algorithm for RAIM resulted in 11 ms, while using also the iterative
algorithm resulted in at least 10 times longer computation that additionally
depends on the bias magnitude.

This investigation clearly indicates that using the non-iterative position calculation method, derived in section 3.4.6, while evaluating the RAIM decision

matrix delivers the best performance in terms of error identification. This is achieved at the cost of only minor increased time effort.

7.2.4. Comparison with state of the art systems

In this section, the localization results achieved during this thesis are compared to the results found in the up-to-date literature. For the comparison to be fair, only systems using impulse radio UWB signals are considered. The experimental results found in the literature and showing the 1D accuracy are not included in the comparison, as no prediction can be done, how would they perform in the 2D or 3D scenario.

Although it was shown in the literature (e.g. [Sym13] and [MHS+07]) that the frequency-modulated continuous-wave systems operating in the ISM band can offer good localization performance (at a level of 5 cm to 20 cm), they are not included in this summary. The reason for this is due to their narrowband nature and hence the lack of multipath fading immunity, which was one of the key motivating factors for this work in the first place.

The results of the experiments performed with indoor IR-UWB localization systems, that were published in the recent years, are summarized in the table 7.2. Besides the essential information about the employed method (column 2), the reported position estimation error (column 6) and the measurement update rate (column 5), additional information regarding the hardware setup and the campaign setup are given. The entries in the table are sorted according to the size of the scenario where the systems were tested.

The first five rows list the small-scale scenarios with a size $< 20\,\mathrm{m}^2$, followed by two medium-scale ones. The localization performance, achieved with the system developed in this thesis, is comparable with others, and after applying the correction algorithms it clearly outperforms them. Most of these systems employ relatively long pulses, occupying $\approx 2\,\mathrm{GHz}$ of bandwidth, where a high-speed oscilloscope was often used as a receiver, giving the opportunity to precisely determine the pulses' time of arrival (peak position). In [PAB+10], a commercial UWB-based localization system from Ubisense was employed, and in [SHSM10] a I/Q receiver with two 8-bit 1.5 Gsps ADCs was used, whereas here the developed system is equipped only with a simple comparator.

The last three rows of table 7.2 contain the information about large-scale experiments. Considering the 2D localization performance, the here developed system is significantly better than the ones found in the literature, with the key advantage over [XML11] of having a fully integrated receiver and a dedicated time measurement unit instead of using a high-speed oscilloscope. Moreover,

Table 7.2.: Performance comparison of the impulse-radio UWB localization systems reported in the literature.

Source	Method	Transmitter	Receiver	Update rate	Reported error / type	Tested range	Remarks
[CNG$^+$12]	TOA 2D	1 ns pulse	oscilloscope	50 kHz	7.4 cm / RMS	$1.8 \times 2.4\,\mathrm{m}^2$ 4 APs	static scenario
[PAB$^+$10]	TDOA 2D	pulse 2.5 GHz BW	Ubisense [Ubi13]	5 Hz	$\approx$ 13 cm / —	$3 \times 3\,\mathrm{m}^2$ 4 APs	moving robot
[ZKCC08]	TDOA 2D	300 ps pulse	—	1 MHz	$\approx$ 11 cm / —	$\sim 13\,\mathrm{m}^2$ per 4 APs	multi-cell
[SKZ08]	TDOA 2D	pulse 2 GHz BW	—	—	15 cm / RMS	$\sim 13\,\mathrm{m}^2$ per 4 APs	multi-cell, moving robot
this work (section 7.1)	TDOA 2D & 3D	250 ps pulse 7 GHz BW	auto-correlation rec. + comparator	1 MHz	< 3 cm / mean 3D	$2.5 \times 6\,\mathrm{m}^2$ 5 APs	static scenario
[ZLGC11]	TDOA/TOA 2D	1 ns pulse f_M=4 GHz	energy detection + oscilloscope	3.2 MHz	1 cm / mean 8 cm / RMS	$4.8 \times 6\,\mathrm{m}^2$ 5 APs	only 2 MU positions
[SHSM10], [SMS12]	TDOA 2D	2 ns pulse f_M=3.5 GHz	QACR 1.5 Gsps	—	$\approx$ 11 cm / RMS	$5 \times 8\,\mathrm{m}^2$ 4 APs	2-way-ranging, moving robot
[MFN$^+$07]	~TDOA 2D	2 ns pulse	—	—	$\approx$ 18 cm / —	$8 \times 8\,\mathrm{m}^2$ 3 APs	1 sync. AP, static scenario
this work (section 7.2)	TDOA 2D & 3D	250 ps pulse 7 GHz BW	auto-correlation rec. + comparator	1 MHz	3.4 cm / mean 2D 11.4 cm / mean 3D 7.32 cm / RMS 2D 18.5 cm / RMS 3D	$9.5 \times 10\,\mathrm{m}^2$ 6 APs	static scenario
[XML11]	T(D)OA[15] 2D	pulse f_M=4 GHz	LNA + oscilloscope	10 MHz	8 cm / RMS	$12 \times 10\,\mathrm{m}^2$ 3 APs	static scenario

[15] During the experiment first the TOA measurements were performed and later subtracted in order to obtain time differences.

despite placing the APs in similar heights, a good 3D localization accuracy has been achieved, a feature which other listed system do not offer.

7.3. Localization reliability improvement using hybrid UWB-IMU system

In this work, it was shown until now, that in indoor scenarios the radio-based (in this case UWB) localization systems can achieve an accuracy in the lower centimeter region. This solution, assuming a certain confidence interval defined by the systems precision, is long-term stable; hence it does not drift away with time in static conditions. All radio technologies have however one obvious drawback, namely the number of access points which are required for a coverage allowing reliable operation. This number varies depending on the scenario, but in general it is considered to be high. This is due to the fact that during the system-deployment phase one has to account for the scenario geometry and ensure line-of-sight conditions between the MU (in the entire region) and a minimum of 3 access points for 2D localization.

Therefore the prospect of using an Inertial Measurement Unit (IMU) for navigating through the scenario, without the need to rely on any infrastructure (reference nodes), is very tempting. However, the pure "inertial" navigation, based on an IMU and additional sensor, such as magnetic compass and a barometer, is subject to drift with time [35] (short-term stable). An interesting approach would be to combine the radio- and the inertial-navigation to profit from their advantages in the optimal way and at the same time to balance-out their drawbacks. The basic idea is to merge the both above described systems in order to achieve a long-term stability thanks to UWB and a smooth trajectory thanks to IMU. The inertial solution would also allow to bridge short periods of time without radio-reception.

In the literature, there are two integration techniques described:

- one option is the so-called *loosely coupled* (LC) integration of both systems. In this case, the localization solution, using only the UWB TDOA measurements, is calculated first and then processed by the fusion filter as a position update. This is the simplest (however not the most efficient) method. Problems occur when not enough APs are available, hence, no stand-alone radio localization solution is possible.

- In this case, the *tightly coupled* (TC) integration of the sensor data is more advantageous [8]. Here, the TDOA and IMU measurements are processed in a Kalman filter, without prior calculation of the localization solution [40]. For more details on the implementation of the LC and TC data fusion in the navigation filter refer to [KAS$^+$11] and [21].

In this section, only the experiment-based proof of concept, regarding the advantages of IMU-UWB integration is presented.

7.3.1. Implementation of the dual UWB-IMU tracking system

The torso-mounted Integrated Pedestrian Navigation System (IPNS) has been developed at the Institute of Systems Optimization (ITE) at KIT. The IPNS platform consists of the IMU, magnetic compass and a barometer. For the calculation of the position solution the step length is estimated based on the accelerometer readings and combined with the heading information of a magnetic compass. The IPNS worn by a person is depicted in Fig. 7.11(b).

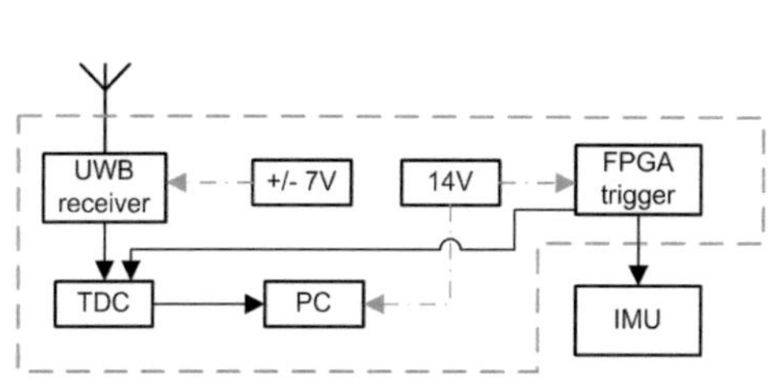

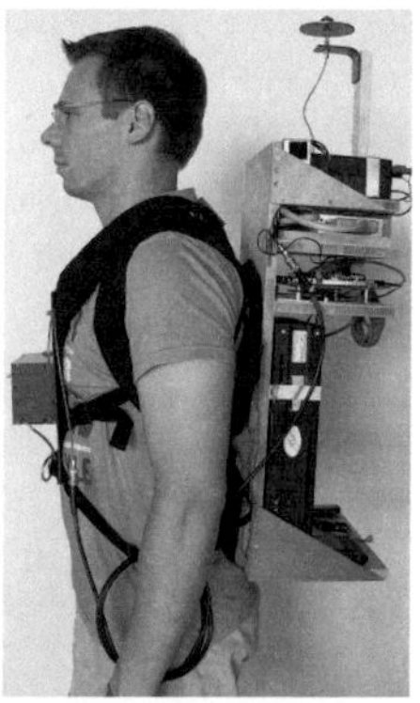

(a) Interconnection scheme of IMU and UWB systems, including synchronization and power supply

(b) The author wearing the hybrid system

Fig. 7.11.: The hybrid UWB-INS localization and tracking system (mobile user part). (a): elements within the grey-dashed box are mounted on a backpack-platform. Black solid arrows indicate synchronization by trigger signals. (b): system ready for measurement.

The UWB localization sub-system consists of the same hardware as used during experiments in sections 7.1 and 7.2. The mobile user wearing the IPNS is equipped with a UWB receiver. Three APs, distributed in the scenario, operate in the Tx mode. The use of three APs allows only the calculation of 2D localization solution from TDOA measurements (reference AP: nr. 1). In

Fig. 7.11(a), the interconnection scheme of the UWB and IPNS is depicted. Both systems were mounted on a wearable platform, called the *Backpack Navigator* (BN). The BN is equipped with the TDC module for measuring the time-differences between the received UWB pulses, a PC in micro-ATX case for collecting the TDOA data from the TDC and batteries for portable operation. The sub-systems are synchronized using a clock signal which has been implemented on an FPGA. The most important parameters of the integrated system are summarized in table 7.3.

Table 7.3.: Parameter of the experimental IMU-UWB system.

Parameter	Value
Synchronization / position update rate	10 Hz
TDOA averaging at TDC	100-fold
Rx UWB antenna height	180 cm
Total weight of the BN	9.5 kg
Battery operation time	1 h
Overall power consumption	$\approx 40\,W$
UWB/IPNS power consumption	$\approx 2\,W$

7.3.2. Experimental results of radio-assisted indoor inertial tracking

The measurement campaign was conducted in an empty room of $6\,m \times 6\,m$ size, with 3 transmitting UWB APs (marked as dots in the below plots). APs 1 and 2 were mounted on a window frame (the reason why they are outside floor outline) and AP 3 was fixed to a shelf. The pure UWB measurements are plotted as crosses in the map. The ground truth of the trajectory is given by the measurement points, which in the plots are connected with dashed lines. The person did not walk precisely on the dashed path in straight lines, but was only stepping the edge points - the resulting trajectory was oval. The color of the trajectory and UWB localization solution is changing over time, according to the legend.

In Fig. 7.12, the UWB APs were not active. This demonstrates how strong the pure inertial tracking solution (strapdown + step length update) drifts away. The problem of the lack of the long term stability of such a system is prominent. In Fig. 7.13 (top), the navigation solution is shown, obtained by the loose integration of the positions provided by the IMU and UWB systems. As expected, the UWB positioning is very stable and the radial spread of the positions originates from the movements of the person wearing the system. Between 30 s to 90 s the AP2 was turned off. During this time, the pure UWB

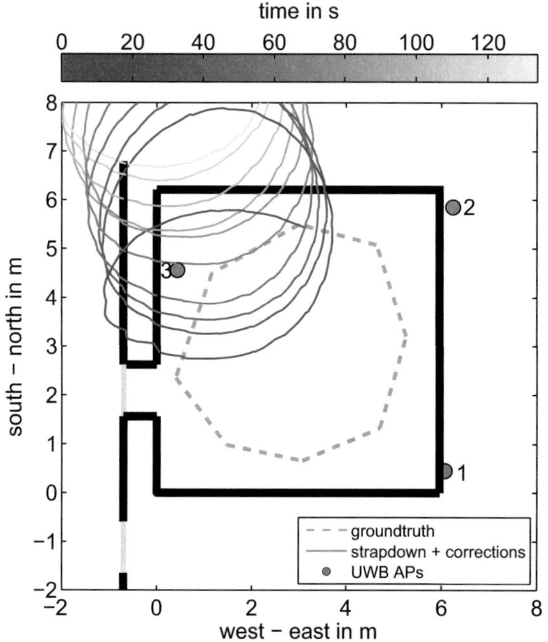

Fig. 7.12.: Solution delivered by the IPNS stand-alone.

solution was not possible, giving no updates to the LC fusion filter, hence the inertial solution drifted away northwards. After 60 s of UWB black-out the position is corrected, as soon as the UWB update is available again.

During the phase of insufficient UWB reception there was only one TDOA measurement obtained (between AP1 and 3). This measurement however still contains some information that can be used to improve the positioning solution. This is possible in the tight integration. The same run, this time with TC fusion, is depicted in Fig. 7.13 (bottom). The $TDOA_{13}$ allows for a correction perpendicular to the time-difference hyperbola. As the person is walking in a circle, the direction of the corrections is changing over time. Although the result is not quite as good as with three APs, the 2D correction is done and the solution does not drift away (despite one outlier). The tightly coupled approach exhibits a certain transient phase in the first 10 s of the measurement and the estimated trajectory is slightly stretched in the west-east direction, however it still benefits even from only one valid TDOA.

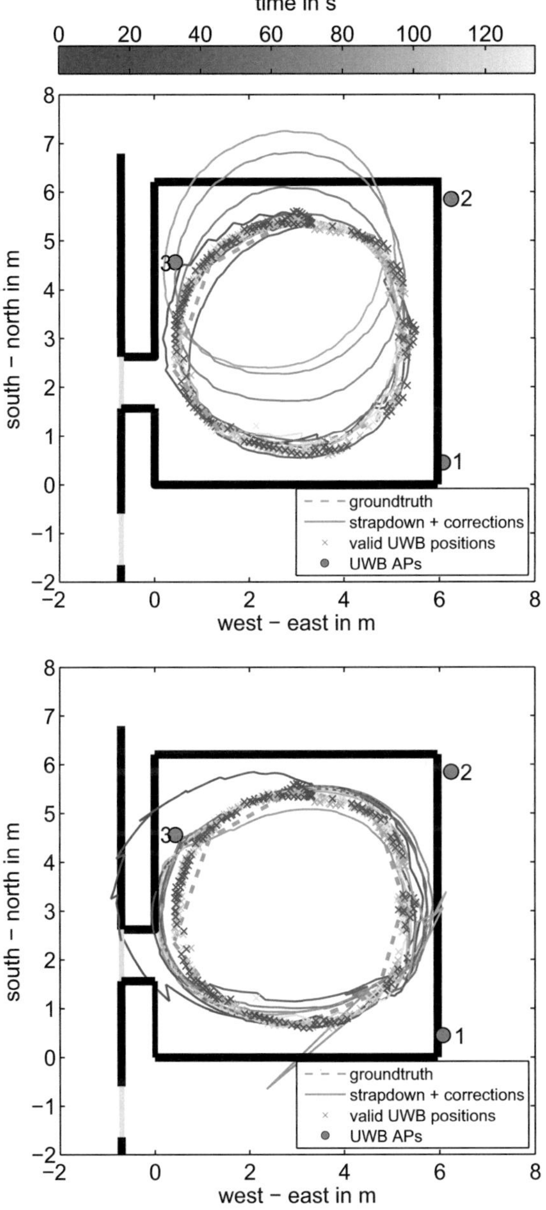

Fig. 7.13.: Tacking results in the case of an AP2 outage between 30 s to 90 s. Top: loosely coupled integration, bottom: tightly coupled integration.

The results presented in this section, have again proven the excellent performance of the designed UWB localization system; this time under dynamic conditions and in real time operation. It was also shown that the potential problem with sufficient radio-coverage in large scenarios, required for UWB stand-alone positioning, can successfully be overcome by the combination with inertial sensors.

8. Conclusions

Nowadays, thanks to smart phones and tablet computers, the uninterrupted wireless access to information has become obvious. In addition, in the past years the clear tendency towards continuous awareness of own position could be observed. The amount of hand-held devices (e.g. cell phones or photo cameras), equipped with GPS receivers sold almost doubles every year. Based on this, it can be easily deduced, that most of the users would not like to resign from this convenience while being indoors. The GPS signal, however, before reaching an indoor receiver, is in the most cases highly attenuated and distorted to an extent that it becomes useless for localization. Moreover, the localization accuracy required in indoor scenarios is several orders of magnitude higher than the one needed to navigate through streets in the city.

The interest for precise indoor localization is high among the potential consumers, but also in the research community. The natural drive is to demand for more - also for more accuracy. The overview presented in chapter 1 shows that a variety of technologies are being examined in this context, where the radio frequency approaches show the best potential. This could enable applications, that are beyond the expectations of today: Wireless localization in challenging environments, e.g. industrial halls? Preferably with decimeter accuracy, or even better? In real time? The impulse-radio Ultra-Wideband technology has this potential: short impulses can deal very well with multipath, do not undergo fading, and thanks to their short time duration offer very good resolution that can be used for localization. The motivation for this work was to design an impulse-based indoor localization system and investigate all its characteristics, which potentially influence the achievable position determination performance.

Criteria and hardware for transceiver architecture comparison
The fundamental question that needs to be answered, is the one about efficient UWB transceiver architecture. This has been handled in section 4, where three receiver structures were investigated in simulation, as well as in the measurements. The self-designed common hardware platform was used to ensure fair comparison, which was never done before in the literature. Based on this, it is possible to state, that for short-range applications, where the SNR is high, the incoherent auto-correlation receiver is the best choice. On the other hand, if

long operation ranges are envisioned, the coherent correlation receiver offers the best signal-to-noise ratio in the received signal. If the optimization for a particular scenario is unwanted, the transmitted-reference architecture offers the best trade-off between complexity and achievable operation range. The UWB transmitter designed in this work operates in compliance with the FCC regulation. The bandwidth of the developed receiver exceeds 8 GHz (130 % relative) and is only limited on purpose by the bandpass filter.

Realization complexity analysis and
applications of IR-UWB correlation receiver
In the following step, special attention has been given to the coherent UWB receiver, as it has the best performance in the above described tests. Unlike the transmitted-reference or auto-correlation scheme, the correlation receiver requires very precise synchronization. This has been investigated in section 5. In this work the synchronization and impulse tracking schemes were developed and for the first time were implemented on an FPGA and tested with UWB hardware. The performed comprehensive parameter study revealed that, different than suspected in theory, it is preferable to use rather low synchronization thresholds and synchronization scheme with medium search step size.

Based on calculations and verification by experiments it has been found that high frequency accuracy of the transmitter and receiver clocks is essential for a good quality of IR-UWB transmission when using the correlation receiver. It has been demonstrated that picosecond-precise and fast-adjustable clock signals can be realized with a phase-locked-loop and an adjacent programmable delay element.

Additionally, in section 5.4, the ranging potential of the correlation receiver was verified. The results obtained indicate an RMS error of only ±2.3 mm in 1D. This makes this system an ideal candidate for two-way-ranging applications. The investigations performed indicate that the use of the correlation receiver in ranging and localization applications is more advisable than for pure data transmission purposes. The reason for this is the high effort (time and hardware) required to obtain and maintain long-term synchronization.

Efficient position estimation and error detection methods
In the further part of the thesis, the focus was put on the localization aspects. For this purpose the time efficient algorithms, that give a possibility of the real time operation, are required. In this work, the modified version of the Bancroft algorithm, that is capable of non-iterative solving of the TDOA localization equations, was introduced (section 3.4). This implementation,

compared with iterative algorithms offers approximately thirty times shorter computation times. Furthermore, it was identified that the combination of the modified Bancroft with Levenberg-Marquardt algorithm offers an optimal trade-off between computation time and precision.

Moreover, the suitable RAIM method, for use with indoor location systems was identified. This approach profits from the fast, non-iterative position estimation method. The performance in terms of the computation time and accuracy of these algorithms have been evaluated with real data in chapter 7. The aspect of the geometrical configuration of access points, and its influence on the localization error was analyzed theoretically and from the practical point of view. This gives a universal guideline for planning of future wireless localization systems.

Subsystems analysis, error modeling and realization limits assessment
Following the logical order of the localization system development, the parameters of the consecutive transmitter and receiver hardware components have been investigated in order to better understand their influence on the achievable accuracy and the precision of the overall system. One of the major issues is the antenna radiation characteristic. As the mobile user moves through the scenario, the relative angles and distances between him and the access points change. To maintain a good accuracy through the scenario, two compensation methods of the angle-dependent antenna radiation characteristic were developed. Both, the iterative and non-iterative versions were successfully tested in practice, improving the localization results by several centimeters.

Furthermore, in chapter 6, the signal detection method based on a comparator and the precise time measurement unit were examined. Two effects, the threshold-trigger offset and the TDOA variance errors were quantified. Corrections methods of both these phenomena have been proposed, which delivered very good results in the experiments. Moreover, based on the error models introduced, the Cramer-Rao lower bound has been derived for this 3D TDOA localization system. The derived CRLB can be used as the benchmark for the localization system precision lower limit. The correctness of this model has been confirmed by measurements.

**Localization accuracy and precision verification
in realistic indoor scenarios**
In order to prove the effectiveness of the above described methods a 3D indoor localization demonstrator, based on impulse-radio UWB TDOA principle has been developed. The measurements conducted in a small-scale scenario have shown the efficiency of the proposed algorithms and resulted in an average

3D positioning error as small as 3 cm. The localization tests performed in the industrial hall have proven the correctness of the accuracy and the precision estimation methods, both depending on the previously described geometrical configuration of the system. The resulting average horizontal and 3D accuracy are 3.4 cm and 11.4 cm, respectively. The comparison with the literature published up to date proves the excellent performance of the localization system developed in this thesis.

In the final section of chapter 7, the issue of radio-coverage and its potential shortage was considered. As a possible solution, for situations where the UWB stand-alone localization is not available, a multi-sensor systems were proposed. The UWB localization system has been successfully integrated with an Inertial Pedestrian Navigation System. With this hybrid platform, a series of measurements in a typical dynamic indoor scenario has been performed. The results demonstrate the high potential of multi-sensor data fusion for indoor positioning and tracking applications, resulting in an improved long- and short-term stability of the entire system.

In this thesis, the impulse UWB localization system has been thoroughly analyzed and the strengths and weaknesses of this approach have been presented. The developed and verified error models as well as given guidelines can be used for design and optimization of future application-tailored positioning systems.

A. Derivation of the Fisher Information Matrix for 3D TDOA localization

In this appendix the derivation of the Fisher Information Matrix (FIM) elements for the 3D TDOA localization case is presented. This is required to calculate the Cramer-Rao Lower Bound described in section 6.5 and later used in chapter 7. The MU position vector r_S includes three coordinates, hence the FIM has the following form:

$$
I(r_S) = \begin{bmatrix} I_{11} & I_{12} & I_{13} \\ I_{21} & I_{22} & I_{23} \\ I_{31} & I_{32} & I_{33} \end{bmatrix} = \begin{bmatrix} I_{xx} & I_{xy} & I_{xz} \\ I_{yx} & I_{yy} & I_{yz} \\ I_{zx} & I_{zy} & I_{zz} \end{bmatrix}. \tag{A.1}
$$

Derivation of the first element of the FIM

The first element of the Fisher Information Matrix is defined as follows:

$$
I_{11} = E\left\{\left(\frac{\partial \ln(p(\Delta \tilde{d}_{ij}|r_S))}{\partial x_S}\right)\left(\frac{\partial \ln(p(\Delta \tilde{d}_{ij}|r_S))}{\partial x_S}\right)\right\}. \tag{A.2}
$$

The first step is to derive the natural logarithm of the conditional joint PDF defined in equation (6.6):

$$
\ln\left(p(\Delta \tilde{d}_{ij}|r_S)\right) = \ln\left(\prod_{i=1;j=2}^{N} \frac{1}{\sqrt{2\pi\sigma_{\Delta\omega}^2}} \exp\left(-\frac{(\Delta \tilde{d}_{ij} - \Delta d_{ij})^2}{2\sigma_{\Delta\omega}^2}\right)\right) \tag{A.3}
$$

$$
= \ln\left(\frac{1}{\sqrt{2\pi\sigma_{\Delta\omega}^2}} \exp\left(-\sum_{i=1;j=2}^{N} \frac{(\Delta \tilde{d}_{ij} - \Delta d_{ij})^2}{2\sigma_{\Delta\omega}^2}\right)\right) \tag{A.4}
$$

$$= \ln\left(\frac{1}{\sqrt{2\pi\sigma_{\Delta\omega}^2}}\right) - \frac{1}{2\sigma_{\Delta\omega}^2} \sum_{i=1;j=2}^{N} (\Delta\tilde{d}_{ij} - \Delta d_{ij})^2. \qquad (A.5)$$

In case when the distance-independent time measurement noise is assumed, the partial derivative of (A.5) with regard to x_S has the following form:

$$\frac{\partial \ln(p(\Delta\tilde{d}_{ij}|r_S))}{\partial x_S} = \underbrace{\frac{\partial}{\partial x_S} \ln\left(\overbrace{\frac{1}{\sqrt{2\pi\sigma_{\Delta\omega}^2}}}^{const}\right)}_{=0}$$

$$- \frac{1}{2\sigma_{\Delta\omega}^2} \sum_{i=1;j=2}^{N} 2(\Delta\tilde{d}_{ij} - \Delta d_{ij})\left(\frac{\partial\Delta\tilde{d}_{ij}}{\partial x_S} - \frac{\partial\Delta d_{ij}}{\partial x_S}\right) \qquad (A.6)$$

$$= -\frac{1}{\sigma_{\Delta\omega}^2} \sum_{i=1;j=2}^{N} (\Delta\tilde{d}_{ij} - \Delta d_{ij})\left(\frac{\partial\Delta\tilde{d}_{ij}}{\partial x_S} - \frac{\partial\Delta d_{ij}}{\partial x_S}\right). \qquad (A.7)$$

In the next step the partial derivative of the estimated distance difference Δd_{ij} with regard to x_S is calculated (second the last term in (A.7)):

$$\frac{\partial\Delta\tilde{d}_{ij}}{\partial x_S} = \underbrace{\frac{\partial}{\partial x_S}\left(\sqrt{(x_S - x_i)^2 + (y_S - y_i)^2 + (z_S - z_i)^2} + w_i\right)}_{D_{x_i}}$$

$$- \underbrace{\frac{\partial}{\partial x_S}\left(\sqrt{(x_S - x_j)^2 + (y_S - y_j)^2 + (z_S - z_j)^2} + w_j\right)}_{D_{x_j}}. \qquad (A.8)$$

As the noise w is considered here to be distance-independent, its derivative is 0. Therefore the term D_{x_i} from (A.8) is calculated as follows:

$$D_{x_i} = \frac{1}{2}\left((x_S - x_i)^2 + (y_S - y_i)^2(z_S - z_i)^2\right)^{-\frac{1}{2}} \qquad (A.9)$$

$$\cdot \left(\frac{\partial}{\partial x_S}(x_S - x_i)^2 + \underbrace{\frac{\partial}{\partial x_S}(y_S - y_i)^2}_{=0} + \underbrace{\frac{\partial}{\partial x_S}(z_S - z_i)^2}_{=0}\right) \qquad (A.10)$$

$$= \frac{(x_S - x_i)}{\sqrt{(x_S - x_i)^2 + (y_S - y_i)^2 + (z_S - z_i)^2}}. \qquad (A.11)$$

By proceeding in the same way with D_{x_j} and substituting both terms back into (A.8) results in:

$$\frac{\partial \Delta \tilde{d}_{ij}}{\partial x_S} = \frac{(x_S - x_i)}{\sqrt{(x_S - x_i)^2 + (y_S - y_i)^2 + (z_S - z_i)^2}} \tag{A.12}$$

$$- \frac{(x_S - x_j)}{\sqrt{(x_S - x_j)^2 + (y_S - y_j)^2 + (z_S - z_j)^2}}. \tag{A.13}$$

Now, having calculated all the terms of (A.7) must have to inserted into (A.2). Moving the constant terms to the front, results in:

$$I_{11} = \left(-\frac{1}{\sigma_{\Delta w}^2}\right)^2 \cdot E\left\{\left(\sum_{i=1;j=2}^{N} (\Delta \tilde{d}_{ij} - \Delta d_{ij})(D_{x_i} - D_{x_j})\right)^2\right\} \tag{A.14}$$

$$= \left(-\frac{1}{\sigma_{\Delta \omega}^2}\right)^2 \cdot \sum_{i=1;j=2}^{N} \left[\underbrace{E\left\{(\Delta \tilde{d}_{ij} - \Delta d_{ij})^2\right\}}_{\text{term ED}} \cdot \underbrace{E\left\{(D_{x_i} - D_{x_j})^2\right\}}_{\text{const}}\right] \tag{A.15}$$

$$= \frac{1}{\sigma_{\Delta \omega}^2} \cdot \sum_{i=1;j=2}^{N} \left(D_{x_i} - D_{x_j}\right)^2. \tag{A.16}$$

The result of $(\Delta \tilde{d}_{ij} - \Delta d_{ij})$ is a pure, normally distributed noise Δw with standard deviation $\sigma_{\Delta w}$. The mean value of the PDF that resulted from a squared normal distribution is $\mu = \sigma_{\Delta w}^2$ Therefore, the term ED from equation (A.15) is equal $\sigma_{\Delta w}^2$.

Other elements of the FIM

Analogous to the derivation presented in section A, the other two elements of the FIM main diagonal are:

$$I_{22} = E\left\{\left(\frac{\partial \ln(p(\Delta \tilde{d}|r_S))}{\partial y_S}\right)\left(\frac{\partial \ln(p(\Delta \tilde{d}|r_S))}{\partial y_S}\right)\right\} = \tag{A.17}$$

$$= \frac{1}{\sigma_{\Delta \omega}^2} \sum_{i=1;j=2}^{N} (D_{y_i} - D_{y_j})^2 \tag{A.18}$$

and

$$I_{33} = \frac{1}{\sigma^2_{\Delta\omega}} \sum_{i=1;j=2}^{N} (D_{z_i} - D_{z_j})^2.$$ (A.19)

For the elements on both sides of the main diagonal the following relations hold: $I_{12} = I_{21}$, $I_{13} = I_{31}$ and $I_{23} = I_{32}$.

$$I_{12} = \frac{1}{\sigma^2_{\Delta w}} \sum_{i=1;j=2}^{N} (D_{x_i} - D_{x_j})(D_{y_i} - D_{y_j})$$ (A.20)

$$I_{13} = \frac{1}{\sigma^2_{\Delta w}} \sum_{i=1;j=2}^{N} (D_{x_i} - D_{x_j})(D_{z_i} - D_{z_j})$$ (A.21)

$$I_{23} = \frac{1}{\sigma^2_{\Delta w}} \sum_{i=1;j=2}^{N} (D_{y_i} - D_{y_j})(D_{z_i} - D_{z_j})$$ (A.22)

Inverse of the FIM and the lower limit of variance

For calculation of the Cramer-Rao lower bound according to (6.9), the inverse of the Fisher Information Matrix has to be calculated first. The inverse of the 3×3 FIM can be obtained using the Cramer's rule:

$$I^{-1} = \begin{bmatrix} I_{11} & I_{12} & I_{13} \\ I_{21} & I_{22} & I_{23} \\ I_{31} & I_{32} & I_{33} \end{bmatrix}^{-1}$$ (A.23)

$$= \frac{1}{det(I)} \begin{bmatrix} I_{22}I_{33} - I_{23}I_{32} & I_{13}I_{32} - I_{12}I_{33} & I_{12}I_{23} - I_{13}I_{22} \\ I_{23}I_{31} - I_{21}I_{33} & I_{11}I_{33} - I_{13}I_{31} & I_{13}I_{21} - I_{11}I_{23} \\ I_{21}I_{32} - I_{22}I_{31} & I_{12}I_{31} - I_{11}I_{32} & I_{11}I_{22} - I_{12}I_{21} \end{bmatrix}.$$ (A.24)

The determinant can easily be using the scheme of Sarrus. The resulting lower bound on the measurement variance (CRLB) of the mobile user position in the 3D TDOA localization system is:

$$\sigma^2_{MU}(\tilde{r}_S) \geq \text{trace}\left(I^{-1}(r_S)\right) = I_{11}^{-1} + I_{22}^{-1} + I_{33}^{-1}.$$ (A.25)

Bibliography

[aca07]   acam-messelectronic gmbh. TDC-GPX: Ultra-high Performance 8 Channel Time-to-Digital Converter. http://www.acam.de/file admin/Download/pdf/English/DB_GPX_e.pdf, Jan. 2007.

[ACH$^+$01] M. Addlesee, R. Curwen, S. Hodges, J. Newman, P. Steggles, A. Ward, and A. Hopper. Implementing a Sentient Computing System. *Computer*, 34(8):50–56, Aug. 2001.

[ADA08]   Y.S. Alj, Ch. Despins, and S. Affes. Impact of multipath interference on the performance of an UWB fast acquisition system for ranging in an indoor wireless channel. In *Communication Networks and Services Research Conference (CSNR)*, pages 390–396, May 2008.

[Aet10]   Aether Wire and Location Inc. UWB RTLS from Aether Wire. http://www.aetherczar.com/?p=705, Jun. 2010.

[Agi09]   Agilent Technology. Advanced Design System (ADS). http://www.home.agilent.com/en/pc-1297113/advanced-design-system-ads, 2009.

[Ana09]   Analog Devices. Ultrafast 3.3V/5V Single-Supply SiGe Comparators: ADCMP572 and ADCMP573. http://www.analog.com/ static/imported-files/data_sheets/ADCMP572_573.pdf, Apr. 2009.

[BAK$^+$02] S. Banerjee, S. Agarwal, K. Kamel, A. Kochut, C. Kommareddy, T. Nadeem, P. Thakkar, Bao Trinh, A. Youssef, M. Youssef, R.L. Larsen, A. Udaya Shankar, and A. Agrawala. Rover: Scalable Location-Aware Computing. *Computer*, 35(10):46–53, Oct. 2002.

[Ban85]   S. Bancroft. An algebraic-solution to the GPS-equations. *IEEE Transactions on Aerospace and Electronic Systems*, 21(1):56–59, Jan. 1985.

[Ber11]   H. Berchtold. Theoretische und messtechnische Untersuchung der Leistungsfähigkeit verschiedener UWB Empfangstechniken im Hinblick auf den Signal-Rausch-Abstand. Bachelor's thesis, Karlsruher Institut für Technologie (KIT), Sep. 2011.

[BH99]   J.D. Bard and F.M. Ham. Time Difference of Arrival Dilution of Precision and Applications. *IEEE Transactions on Signal Processing*, 47(2):521–523, Feb. 1999.

[BHY$^+$02] H. Bing, X.D. Hou, X. Yang, T. Yang, and C. Li. A "Two-Step" Synchronous Sliding Method of Sub-Nanosecond Pulses for Ultra-Wideband (UWB) Radio. In *IEEE International Conference on Communications, Circuits and Systems*, volume 1, pages 142–145, Jun. 2002.

[BP00]   P. Bahl and V.N. Padmanabhan. RADAR: An In-Building RF-based User Location and Tracking System. In *19th Annual Joint Conference of the IEEE Computer and Communications Societies*, volume 2, pages 775–784, Mar. 2000.

[BZS$^+$05] S. Bagga, Lujun Zhang, W.A. Serdijin, J.R. Long, and E.B. Busking. A Quantized Analog Delay for an ir-UWB Quadrature Downconversion Autocorrelation Receiver. In *IEEE International Conference on Ultra-Wideband (ICU)*, pages 328–332, Sep. 2005.

[CA94]   J. Chaffee and J. Abel. GDOP and the Cramer-Rao bound. In *IEEE Position Location and Navigation Symposium*, pages 663–668, Apr. 1994.

[CB07]   Y. Chen and N.C. Beaulieu. SNR Estimation Methods for UWB Systems. *IEEE Transactions on Wireless Communications*, 6(10):3836–3845, Oct. 2007.

[CBG99]   T. Coleman, M.-A. Branch, and A. Grace. *Optimization Toolbox for use with MATLAB*. The Mathworks, Inc., 2nd edition edition, 1999.

[CCT00]   A.R. Conn, N.I.M. Could, and P.L. Toint. *Trust-Region Methods*. Society for Industrial and Applied Mathematics and Mathematical Programming Society, 2000.

[CDW08]   A. Conti, D. Dardari, and M.Z. Win. Experimental Results on Cooperative UWB Based Positioning Systems. In *IEEE International Conference on Ultra-Wideband (ICUWB)*, volume 1, pages 191–195, 2008.

[CH94]   Y.T. Chan and K.C. Ho. A Simple and Efficient Estimator for Hyperbolic Location. *IEEE Transactions on Signal Processing*, 42(8):1905–1915, Aug. 1994.

[Che11]   H.-C. Chen. Design Approach of a Low-Jitter DDS-Like Frequency Synthesizer Using Mixed-Mode Signal Processing. In *First International Conference on Robot, Vision and Signal Processing (RVSP)*, pages 336–339, Nov. 2011.

[Chr09]   B. Christianson. Global Convergence using De-linked Goldstein or Wolfe Linesearch Conditions. *Advanced Modeling and Optimization (AMO)*, 11(1):25–31, 2009.

[CNG$^+$12] O. Cetin, H. Nazli, R. Gurcan, H. Ozturk, H. Guneren, Y. Yelkovan, M. Cayir, H. Celebi, and H.P. Partal. An Experimental Study of High Precision TOA based UWB Positioning Systems. In *IEEE International Conference on Ultra-Wideband (ICUWB)*, pages 357–361, Sep. 2012.

[Coh68]   S.B. Cohn. A Class of Broadband Three-Port TEM-Mode Hybrids. *IEEE Transactions on Microwave Theory and Techniques*, 16(2):110–116, Feb. 1968.

[Col08]   S. Colwell. UWB Location Tech on a Roll. *GPS World*, 19(4):34–35, Apr. 2008.

[Com12]   Computer Simulation Technology. CST Microwave Studio. http://www.cst.com/, 2012.

[CP09]   J.-C.S. Chieh and A.-V. Pham. Development of a broadband Wilkinson power combiner on Liquid Crystal Polymer. In *Asia Pacific Microwave Conference (APMC)*, pages 2068–2071, Dec. 2009.

[CZ07]   Z. Chen and Y. Zhang. A modified synchronization scheme for impulse-based UWB. In *6th International Conference on Information, Communications Signal Processing*, pages 1–5, Dec. 2007.

[DB05]   G. Durisi and S. Benedetto. Comparison between Coherent and Noncoherent Receivers for UWB Communications. *EURASIP Journal on Applied Signal Processing*, 2005:359–368, Jan. 2005.

[DDGC11] A. De Angelis, M. Dionigi, R. Giglietti, and P. Carbone. Experimental Comparison of Low-Cost Sub-Nanosecond Pulse Generators. *IEEE Transactions on Instrumentation and Measurement*, 60(1):310–318, Jan. 2011.

[DDM⁺09] A. De Angelis, M. Dionigi, A. Moschitta, R. Giglietti, and P. Carbone. Characterization and Modeling of an Experimental UWB Pulse-Based Distance Measurement System. *IEEE Transactions on Instrumentation and Measurement*, 58(5):1479–1486, May 2009.

[DR08] W. Dahmen and A. Reusken. *Numerik für Ingenieure und Naturwissenschaftler*, volume 2., korrigierte Aufl. Springer Verlag GmbH, 2008.

[DSD⁺07] J. Dederer, B. Schleicher, F. De Andrade Tabarani Santos, A. Trasser, and H. Schumacher. FCC compliant 3.1-10.6 GHz UWB Pulse Radar System Using Correlation Detection. In *IEEE/MTT-S International Microwave Symposium (IMS)*, pages 1471–1474, Jun. 2007.

[DST⁺08] J. Dederer, B. Schleicher, A. Trasser, T. Feger, and H. Schumacher. A fully monolithic 3.1-10.6 GHz UWB Si/SiGe HBT impulse-UWB correlation receiver. In *IEEE International Conference on Ultra-Wideband (ICUWB)*, volume 1, pages 33–36, Sep. 2008.

[DTS06a] J. Dederer, A. Trasser, and H. Schumacher. Compact SiGe HBT Low Noise Amplifiers for 3.1-10.6 GHz Ultra-Wideband Applications. In *IEEE Topical Meeting on Silicon Monolithic Integrated Circuits in RF Systems*, pages 391–394, Jan. 2006.

[DTS06b] J. Dederer, A. Trasser, and H. Schumacher. SiGe Impulse Generator for Single-Band Ultra-Wideband Applications. In *Third International SiGe Technology and Device Meeting (ISTDM)*, pages 1–2, May 2006.

[Eis06] M. Eisenacher. *Optimierung von Ultra-Wideband-Signalen (UWB)*. Doktorarbeit, Universität Karlsruhe (TH), 2006.

[Eka13] Ekahau. Ekahau Real Time Location System: Wi-Fi Based Asset Management and People Tracking Solution. http://www.ekahau.com/products/real-time-location-system/overview.html, Jan. 2013.

[Ele06]   Electronic Communications Committee. ECC Decision of 24 March 2006 on the harmonised conditions for devices using Ultra-Wideband (UWB) technology in bands below 10.6 GHz. Mar. 2006.

[EQK$^+$12]   E. Elkhouly, Liu Quanhua, M. Kuhn, M. Mahfouz, and A. Fathy. A Study of the Effect of Jitter on UWB System Positioning Accuracy. In *IEEE Radio and Wireless Symposium (RWS)*, pages 287–290, Jan. 2012.

[EZL$^+$07]   A. Elsherbini, C. Zhang, S. Lin, M. Kuhn, A. Kamel, A.E. Fathy, and H. Elhennawy. UWB antipodal vivaldi antennas with protruded dielectric rods for higher gain, symmetric patterns and minimal phase center variations. In *IEEE Antennas and Propagation Society International Symposium (AP-S)*, pages 1973–1976, Jun. 2007.

[Fed02]   Federal Communications Commission. Revision of Part 15 of the Commission's Rules Regarding Ultra-Wideband Transmission Systems *First Report and Order*, Feb. 2002.

[FGW02]   A. Forsgren, P.E. Gill, and M.H. Wright. Interior Methods for Nonlinear Optimization. *Society for Industrial and Applied Mathematics Review*, 44(4):525–597, Oct. 2002.

[FKML10]   G. Fischer, O. Klymenko, D. Martynenko, and H. Luediger. An Impulse Radio UWB Transceiver with High-Precision TOA Measurement Unit. In *International Conference on Indoor Positioning and Indoor Navigation (IPIN)*, pages 1–8, Sep. 2010.

[GA08]   M.S. Grewal and A.P. Andrews. *Kalman Filtering - Theory and Practice using MATLAB*. John Wiley & Sons, Inc, 3rd edition, 2008.

[GBM10]   H. Gao, P. Baltus, and Q. Meng. 2GSPS 6-bit ADC for UWB Receivers. In *International Symposium on Signals Systems and Electronics (ISSSE)*, volume 1, pages 1–4, Sep. 2010.

[GP09]   S. Gezici and H.V. Poor. Position Estimation via Ultra-Wide-Band Signals. *Proceedings of the IEEE*, 97(2):386–403, Feb. 2009.

[Ham94]   J.D. Hamilton. *Time Series Analysis*. Princeton University Press, 1994.

[HB06]   M. Hanke-Bourgeois. *Grundlagen der numerischen Mathematik und des wissenschaftlichen Rechnens*. Wiesbaden: Teubner, 2., berarb. und erw. aufl. edition, 2006.

[HK08]   K. Hamaguchi and R. Kohno. Development of Experimental TDOA System Test-Bed for Indoor Applications. In *IEEE International Conference on Ultra-Wideband (ICUWB)*, volume 3, pages 201–204, Sep. 2008.

[HS02]   E.A. Homier and R.A. Scholtz. Rapid Acquisition of Ultra-Wideband Signals in the Dense Multipath Channel. In *IEEE Conference on Ultra Wideband Systems and Technologies*, pages 105–109, May 2002.

[HT02]   R. Hoctor and H. Tomlinson. Delay-Hopped Transmitted-Reference RF Communications. In *IEEE Conference on Ultra Wideband Systems and Technologies*, pages 265–269, May 2002.

[JDW08]   D. Jourdan, D. Dardari, and M. Win. Position Error Bound for UWB Localization in Dense Cluttered Environments. *IEEE Transactions on Aerospace and Electronic Systems*, 44(2):613–628, Apr. 2008.

[JMK06]   J.-P. Jansson, A. Mantyniemi, and J. Kostamovaara. A CMOS Time-to-Digital Converter With Better Than 10 ps Single-Shot Precision. *IEEE Journal of Solid-State Circuits*, 41(6):1286–1296, Jun. 2006.

[KAS⁺11]   M. Kleinert, C. Ascher, S. Schleith, G.F. Trommer, and U. Stilla. A New Pedestrian Navigation Algorithm Based on the Stochastic Cloning Technique for Kalman Filtering. In *International Technical Meeting of the Institute of Navigation (ION-ITM)*, pages 662–669, Jan. 2011.

[KMK11]   P. Keranen, K. Maatta, and J. Kostamovaara. Wide-Range Time-to-Digital Converter With 1-ps Single-Shot Precision. *IEEE Transactions on Instrumentation and Measurement*, 60(9):3162–3172, Sep. 2011.

[Kol11]   J. Kolakowski. Application of Ultra-Fast Comparator for UWB Pulse Time of Arrival Measurement. In *IEEE International Conference on Ultra-Wideband (ICUWB)*, pages 470–473, Sep. 2011.

[KPJ04]   H. Kim, D. Park, and Y. Joo.   All-Digital Low-Power CMOS Pulse Generator for UWB System. *Electronics Letters*, 40(24):1534–1535, Nov. 2004.

[Kro12]   J. Kroop. Design von LTCC basierten Packages für den modularen Aufbau von UWB Transceiver. Bachelor's thesis, Karlsruher Institut für Technologie (KIT), Jul. 2012.

[Kun04]   M. Kunert. Peak-Average Power Limitation Criteria for Impulse UWB Devices. Technical report, Siemens VDO Automotive AG, Mar. 2004.

[Lev00]   N. Levanon. Lowest GDOP in 2-D scenarios. *IEE Proceedings - Radar, Sonar and Navigation*, 147(3):149–155, Jun. 2000.

[LHB10]   W. Li, L. Huang, and X. Bai. A pseudo-2bit 4-GSps Flash ADC in 0.18 µm CMOS for an IR-UWB Communication System. In *IEEE International Conference on Ultra-Wideband (ICUWB)*, volume 2, pages 1–4, Sep. 2010.

[LLB84]   C.Q. Li, S.H. Li, and R.G. Bosisio.   CAD/CAE Design of an Improved, Wideband Wilkinson Power Divider. *Microwave Journal*, pages 125–135, Nov. 1984.

[LS02]   J. Y. Lee and R. A. Scholtz. Ranging in a dense multipath environment using an UWB radio link. *IEEE Journal on Selected Areas In Communications*, 20(9):1677–1683, Dec. 2002.

[LS09]   K.W.K. Lui and H. C. So. A study of two-dimensional sensor placement using time-difference-of-arrival measurements. *Digital Signal Processing*, 19(4):650–659, Jul. 2009.

[LSG+09]   M. Leib, E. Schmitt, A. Gronau, J. Dederer, B. Schleicher, H. Schumacher, and W. Menzel. A Compact Ultra-Wideband Radar for Medical Applications. *FREQUENZ*, 63(1-2):1–8, Feb. 2009.

[MÖ8]   R. Mönikes.   *Verwendung differentieller GNSS-Trägerphasenmessungen zur integrierten hochgenauen Positionierung.* Doktorarbeit, Universität Karlsruhe (TH), 2008.

[MFN$^+$07] K. Mizugaki, R. Fujiwara, T. Nakagawa, G. Ono, T. Norimatsu, T. Terada, M. Miyazaki, Y. Ogata, A. Maeki, S. Kobayashi, N. Koshizuka, and K. Sakamura. Accurate Wireless Location/Communication System With 22-cm Error Using UWB-IR. In *IEEE Radio and Wireless Symposium*, pages 455–458, Jan. 2007.

[MH06] C.L. Matson and A. Haji. Biased Cramer-Rao lower bound calculations for inequality-constrained estimators. *Journal of the Optical Society of America*, 23(11):2702–2713, Nov. 2006.

[MHS$^+$07] R. Mosshammer, M. Huemer, R. Szumny, K. Kurek, J. Huttner, and R. Gierlich. A 5.8 GHz Local Positioning and Communication System. In *IEEE/MTT-S International Microwave Symposium (IMS)*, pages 1237–1240, Jun. 2007.

[Mic11] Micrel. SY89295U: 2.5V/3.3V 1.5GHz Precision LVPECL Programmable Delay. http://www.micrel.com/_PDF/HBW/sy89295u.pdf, Mar. 2011.

[MOK09] G. MacGougan, K. O'Keefe, and R. Klukas. Ultra-Wideband Ranging Precision and Accuracy. *Measurement Science & Technology*, 20(9):1–13, Sep. 2009.

[MP06] Ch. Mensing and S. Plass. Positioning algorithms for cellular networks using TDOA. In *31st IEEE International Conference on Acoustics, Speech and Signal Processing (ICASSP)*, pages 4183–4186, May 2006.

[MS08] A. Mohamed and L. Shafai. Investigation on the phase centre of ultra wideband circular monopole antennas. In *IEEE Antennas and Propagation Society International Symposium (AP-S)*, pages 1–4, Jul. 2008.

[MZH08] R. Miri, L. Zhou, and P. Heydari. Timing Synchronization in Rmpulse-Radio UWB: Trends and Challenges. In *Joint IEEE Northeast Workshop on Circuits and Systems and TAISA Conference (NEWCAS-TAISA)*, pages 221–224, Jun. 2008.

[MZM$^+$08] M.R. Mahfouz, C. Zhang, B.C. Merkl, M.J. Kuhn, and A.E. Fathy. Investigation of High-Accuracy Indoor 3-D Positioning Using UWB Technology. *IEEE Transactions on Microwave Theory and Techniques*, 56(6):1316–1330, Jun. 2008.

[Nee09] J. Neering. *Optimization and Estimation Techniques for Passive Acoustic Source Localization*. PhD thesis, Mines ParisTech, Apr. 2009.

[Nor13] Northern Digital Inc. Website. http://www.ndigital.com/life sciences/certus-techspecs.php, Jan. 2013.

[NW05] J. Nocedal and S. Wright. *Numerical Optimization*, volume 2nd Edition. Berlin: Springer New York, 2005.

[PAB$^+$10] A. Prorok, A. Arfire, A. Bahr, J.R. Farserotu, and A. Martinoli. Indoor Navigation Research with the Khepera III Mobile Robot: An Experimental Baseline with a Case-Study on Ultra-Wideband Positioning. In *International Conference on Indoor Positioning and Indoor Navigation (IPIN)*, pages 1–9, Sep. 2010.

[PAK$^+$05] N. Patwari, J.N. Ash, S. Kyperountas, A.O. Hero, R.L. Moses, and N.S. Correal. Locating the Nodes: Cooperative localization in wireless sensor networks. *IEEE Signal Processing Magazine*, 22(4):54–69, Jul. 2005.

[PKC02] P. Prasithsangaree, P. Krishnamurthy, and P. Chrysanthis. On indoor position location with wireless LANs. In *13th IEEE International Symposium on Personal, Indoor and Mobile Radio Communications*, volume 2, pages 720–724, Sep. 2002.

[PLU12] PLUS Location Systems. PLUS Real-Time Location System. http://plus-ls.com/wp-content/uploads/2012/09/plus-system.pdf, Sep. 2012.

[Pri05] N. B. Priyantha. *The Cricket Indoor Location System*. PhD thesis, Massachusetts Institute of Technology (MIT), Jun. 2005.

[PTZW09] E. Pancera, J. Timmermann, T. Zwick, and W. Wiesbeck. Signal Optimization in UWB Systems. In *German Microwave Conference (GeMiC)*, pages 1–4, Mar. 2009.

[PZW11] E. Pancera, T. Zwick, and W. Wiesbeck. Spherical Fidelity Patterns of UWB Antennas. *IEEE Transactions on Antennas and Propagation*, 59(6):2111–2119, Jun. 2011.

[QW05] T.Q.S. Quek and M.Z. Win. Analysis of UWB Transmitted-Reference Communication Systems in Dense Multipath Channels. *IEEE Journal on Selected Areas in Communications*, 23(9):1863–1874, Sep. 2005.

[RB04]   P. Rulikowski and J. Barrett.   Truly Balanced Step Recovery
           Diode Pulse Generator with Single Power Supply. In *IEEE Radio
           and Wireless Conference*, pages 347–350, Sep. 2004.

[Ree05]   J.H. Reed. *An Introduction to Ultra Wideband Communication
           Systems*. Prentice Hall Press, Upper Saddle River, NJ, USA, first
           edition, 2005.

[RFSL04]   S. Roy, J.R. Foerster, V.S. Somayazulu, and D.G. Leeper.
           Ultrawideband Radio Design: The Promise of High-Speed,
           Short-Range Wireless Connectivity. *Proceedings of the IEEE*,
           92(2):295–311, Feb. 2004.

[RGV08]   S. Roehr, P. Gulden, and M. Vossiek.   Precise Distance and
           Velocity Measurement for Real Time Locating in Multipath
           Environments Using a Frequency-Modulated Continuous-Wave
           Secondary Radar Approach. *IEEE Transactions on Microwave
           Theory and Techniques*, 56(10):2329–2339, Oct. 2008.

[RZ75]   H.-M. Rein and M. Zahn. Subnanosecond-Pulse Generator with
           Variable Pulsewidth Using Avalanche Transistors. *Electronics
           Letters*, 11(1):21–23, Sep. 1975.

[SÖ7]   W. Sörgel.   *Charakterisierung von Antennen für die Ultra-
           Wideband-Technik*.   Doktorarbeit, Universität Karlsruhe (TH),
           2007.

[SAT03]   H. Sairo, D. Akopian, and J. Takala. Weighted dilution of preci-
           sion as quality measure in satellite positioning. *IEE Proceedings
           on Radar, Sonar and Navigation*, 150(6):430–436, Dec. 2003.

[Sch10]   U. Schwesinger.   Modellierung eines UWB-Trackingsystems
           zur Stützung von inertialsensorbasierten Navigationssystemen in
           Gebäuden.   Diplomarbeit, Karlsruher Institut für Technologie
           (KIT), May 2010.

[Sch11]   J. Schlichenmaier. Konzept, Aufbau und Analyse eines UWB-
           basierten Lokalisierungssystems. Studienarbeit, Karlsruhe Insti-
           tut für Technologie (KIT), Nov. 2011.

[SDS09]   B. Schleicher, J. Dederer, and H. Schumacher. Si/SiGe HBT
           UWB Impulse Generators with Sleep-Mode Targeting the FCC
           Masks. In *International Conference on Ultra-Wideband (ICU-
           WB)*, pages 674–678, Sep. 2009.

[SDST07]   H. Schumacher, J. Dederer, B. Schleicher, and A. Trasser. Si/SiGe Integrated Circuits for Impulse-Radio UWB Sensing and Communications. In *Asia-Pacific Microwave Conference (APMC)*, pages 1–4, Dec. 2007.

[SGAR07]   P.K. Sagiraju, P. Gali, D. Akopian, and G.V.S. Raju. Enhancing Security in Wireless Networks Using Positioning Techniques. In *IEEE International Conference on System of Systems Engineering (SoSE)*, pages 1–6, Apr. 2007.

[SGK05]   J. Schroeder, S. Galler, and K. Kyamakya. A Low-Cost Experimental Ultra-Wideband Positioning System. In *IEEE International Conference on Ultra-Wideband (ICUWB)*, pages 632–637, Sep. 2005.

[SGKJ06]   J. Schroeder, S. Galler, K. Kyamakya, and K. Jobmann. Practical Considerations of Optimal Three-Dimensional Indoor Localization. In *IEEE International Conference on Multisensor Fusion and Integration for Intelligent Systems*, pages 439–443, Sep. 2006.

[SGKJ07]   J. Schroeder, S. Galler, K. Kyamakya, and K. Jobmann. NLOS Detection Algorithms for Ultra-Wideband Localization. In *4th Workshop on Positioning Navigation and Communication (WPNC)*, pages 159–166, Mar. 2007.

[SGTS09]   B. Schleicher, H. Ghaleb, A. Trasser, and H. Schumacher. FM over impulse radio UWB. In *IEEE International Conference on Ultra-Wideband (ICUWB)*, pages 200–204, Sep. 2009.

[SHH+10]   J. Sachs, M. Helbig, R. Herrmann, M. Kmec, and K. Schilling. On the Range Precision of UWB Radar Sensors. In *11th International Radar Symposium (IRS)*, pages 1–4, Jun. 2010.

[SHSM10]   M. Segura, H. Hashemi, C. Sisterna, and V. Mut. Experimental Demonstration of Self-Localized Ultra Wideband Indoor Mobile Robot Navigation System. In *International Conference on Indoor Positioning and Indoor Navigation (IPIN)*, pages 1–9, Sep. 2010.

[Sit09]   Y.L. Sit. Design and Analysis of an Ultra-Wideband Transmitter. Master's thesis, Universität Karlsruhe (TH), Nov. 2009.

[SKZ08]   P. Sharma, S. Krishnan, and G. Zhang. A Multi-cell UWB-IR Localization System for Robot Navigation. In *IEEE Radio and Wireless Symposium*, pages 851–854, Jan. 2008.

[SMS12]   M. Segura, V. Mut, and C. Sisterna. Ultra Wideband Indoor Navigation System. *Radar, Sonar and Navigation, IET*, 6(5):402–411, Jun. 2012.

[Sto08]   L. Stoica. *Non-Coherent Energy Detection Transceivers for Ultra Wideband Impulse Radio Systems*. PhD thesis, University of Oulu, Feb. 2008. ISBN: 978-951-42-8717-6.

[Str03]   S. Stroh. Ultra-Wideband: Multimedia Unplugged. *IEEE Spectrum*, 40(9):23–27, Sep. 2003.

[SYG09]   I. Sharp, K. Yu, and Y.J. Guo. GDOP Analysis for Positioning System Design. *IEEE Transactions on Vehicular Technology*, 58(7):3371–3382, Sep. 2009.

[Sym13]   Symeo GmbH. Highly Available Position Detection. http://www.symeo.com/cms/upload/PDF/Datasheet_LPR-2D.pdf, Feb. 2013.

[Tha06]   Thales Research and Technology UK. Ultra Wide-Band (UWB) In-Building Positioning Demonstrator. http://telecom.esa.int/telecom/www/object/index.cfm?fobjectid =2803, Jul. 2006.

[TII+06]   A. Tamtrakarn, H. Ishikuro, K. Ishida, M. Takamiya, and T. Sakurai. A 1-V 299μW Flashing UWB Transceiver Based on Double Thresholding Scheme. In *Symposium on VLSI Circuits*, pages 202–203, Jun. 2006.

[Tim10]   J. Timmermann. *Systemanalyse und Optimierung der Ultrabreitband-Übertragung*. Doktorarbeit, Karlsruher Institut für Technologie (KIT), 2010. ISBN: 978-3-86644-460-7.

[TLL+09]   T. Thiasiriphet, M. Leib, D. Lin, B. Schleicher, J. Lindner, W. Menzel, and H. Schumacher. Investigations on a Comb Filter Approach for IR-UWB Systems. *FREQUENZ*, 63(1-2):1–4, Feb. 2009.

[TYM+06]   T. Terada, S. Yoshizumi, M. Muqsith, Y. Sanada, and T. Kuroda. A CMOS Ultra-Wideband Impulse Radio Transceiver for 1-Mb/s Data Communications and ±2.5-cm Range Finding. *IEEE Journal of Solid-State Circuits*, 41(4):891–898, Apr. 2006.

[Ubi13]   Ubisense. Ubisense Real-time Location Systems (RTLS). http://www.ubisense.net/en/rtls-solutions/, Feb. 2013.

[Van00]   J. Vankka. *Direct Digital Synthesizers: Theory, Design and Applications*. PhD thesis, Helsinki University of Technology (TKK), Nov. 2000. ISBN: 951-22-5318-6.

[VD08]   M. Verhelst and W. Dehaene. Analysis of the QAC IR-UWB Receiver for Low Energy, Low Data-Rate Communication. *IEEE Transactions on Circuits and Systems I: Regular Papers*, 55(8):2423–2432, Sep. 2008.

[WAS09]   W. Wiesbeck, G. Adamiuk, and C. Sturm. Basic Properties and Design Principles of UWB Antennas. *Proceedings of the IEEE*, 97(2):372–385, Feb. 2009.

[Wei04]   Z. Weissman. Indoor Location. White Paper, Nov. 2004.

[Whe13]   WhereNet. Website. http://www.zebra.com/us/en/solutions/technology-need/wherenet.html, Jan. 2013.

[WJB$^+$06]   Q. Wu, B.-S. Jin, L. Bian, Y.-M. Wu, and L.-W. Li. An approach to the determination of the phase center of Vivaldi-based UWB antenna. In *IEEE Antennas and Propagation Society International Symposium*, pages 563–566, Jul. 2006.

[WO06]   T. Wada and H. Okada. Transmission Performance of UWB-IR System Using DLL Timing Synchronization. In *Asia-Pacific Microwave Conference (APMC)*, pages 1505–1508, Dec. 2006.

[WS97]   M.Z. Win and R.A. Scholtz. Energy Capture vs. Correlator Resources in Ultra-Wide Bandwidth Indoor Wireless Communications Channels. In *Military Communications Conference (MILCOM)*, volume 3, pages 1277–1281, Nov. 1997.

[WWG08]   B. Waldmann, R. Weigel, and P. Gulden. Method for High Precision Local Positioning Radar Using an Ultra Wideband Technique. In *IEEE/MTT-S International Microwave Symposium (IMS)*, pages 117–120, Jun. 2008.

[XML06]   J. Xu, M. Ma, and C.L. Law. Position Estimation Using UWB TDOA Measurements. In *IEEE International Conference on Ultra-Wideband (ICUWB)*, pages 605–610, Sep. 2006.

[XML08]   J. Xu, M. Ma, and C.L. Law. Position estimation using ultra-wideband time difference of arrival measurements. *IET Science, Measurement Technology*, 2(1):53–58, Jan. 2008.

[XML11]   J. Xu, M. Ma, and C.L. Law. Performance of time-difference-of-arrival ultra wideband indoor localisation. *IET Science, Measurement Technology*, 5(2):46–53, Mar. 2011.

[YG03]   L. Yang and G.B. Giannakis. Low-Complexity Training for Rapid Timing Acquisition in Ultra Wideband Communications. In *IEEE Global Telecommunications Conference (GLOBECOM)*, volume 2, pages 769–773, Dec. 2003.

[YL10]   R. Ye and H. Liu. UWB TDOA Localization System: Receiver Configuration Analysis. In *International Symposium on Signals Systems and Electronics (ISSSE)*, pages 1–4, Sep. 2010.

[YYR+06]   M. Youssef, A. Youssef, Ch. Rieger, U. Shankar, and A. Agrawala. PinPoint: An Asynchronous Time-Based Location Determination System. In *4th International Conference on Mobile Systems, Applications and Services (MobiSys)*, pages 165–176, Jun. 2006.

[Zeb13]   Zebra Enterprise Solutions.   Sapphire DART Hub and Receivers.   http://www.zebra.com/gb/en/solutions/technology-need/uwb-solutions.html, Mar. 2013.

[ZKCC08]   G. Zhang, S. Krishnan, F. Chin, and K.C. Chung. UWB Multicell Indoor Localization Experiment System with Adaptive TDOA Combination. In *IEEE 68th Vehicular Technology Conference (VTC Fall)*, pages 1–5, Sep. 2008.

[ZKM+10]   C. Zhang, M.J. Kuhn, B.C. Merkl, A.E. Fathy, and M.R. Mahfouz. Real-Time Noncoherent UWB Positioning Radar With Millimeter Range Accuracy: Theory and Experiment. *IEEE Transactions on Microwave Theory and Techniques*, 58(1):9–20, Jan. 2010.

[ZL08]   W. Zhang and J. Lin. UWB Impulse Radio Signal Detection with High-Speed Integrated Comparator. *Information Technology Journal*, 7(3):516–521, 2008.

[ZLGC11]   Y. Zhou, C.L. Law, Y.L. Guan, and F. Chin. Indoor Elliptical Localization Based on Asynchronous UWB Range Measurement. *IEEE Transactions on Instrumentation and Measurement*, 60(1):248–257, Jan. 2011.

Own publications

[1] L. Zwirello, J. Timmermann, G. Adamiuk, and T. Zwick. Using Periodic Template Signals for Rapid Synchronization of UWB Correlation Receivers. In *COST 2100, TD(09)848*, May 2009.

[2] L. Zwirello, A. Adamiuk, W. Wiesbeck, and T. Zwick. Measurement Verification of Dual-Orthogonal Polarized UWB Monopulse Radar System. In *IEEE International Conference on Ultra-Wideband (ICUWB)*, pages 1–4, Sep. 2010.

[3] L. Zwirello, M. Janson, C. Ascher, U. Schwesinger, G.F. Trommer, and T. Zwick. Accuracy Considerations of UWB Localization Systems Dedicated to Large-Scale Applications. In *International Conference on Indoor Positioning and Indoor Navigation (IPIN)*, pages 1–5, Sep. 2010.

[4] L. Zwirello, M. Janson, C. Ascher, U. Schwesinger, G.F. Trommer, and T. Zwick. Localization in Industrial Halls via Ultra-Wideband Signals. In *7th Workshop on Positioning Navigation and Communication (WP-NC)*, pages 144–149, Mar. 2010.

[5] L. Zwirello, M. Janson, and T. Zwick. Ultra-Wideband Based Positioning System for Applications in Industrial Environments. In *European Wireless Technology Conference (EuWIT)*, pages 165–168, Sep. 2010.

[6] L. Zwirello, E. Pancera, X. Li, and T. Zwick. Ultra-Wideband Localization Systems - Possibilities, Challenges and Solutions. In *5th Management Comitee/Working Group Meeting and Workshop of COST ic0803*, Nov. 2010.

[7] L. Zwirello, Y.L. Sit, G. Adamiuk, E. Pancera, and T. Zwick. Using the antenna matching characteristic for pulse shaping in IR-UWB transmission. In *4th Management Comitee/Working Group Meeting and Workshop of COST ic0803*, Feb. 2010.

[8] L. Zwirello, C. Ascher, G.F. Trommer, and T. Zwick. Study on UWB/INS Integration Techniques. In *8th Workshop on Positioning Navigation and Communication (WPNC)*, pages 13–17, Apr. 2011.

[9] L. Zwirello, M. Harter, Ch. Heine, and T. Zwick. Analysis and Practical Performance Validation of a Correlation-based UWB Transceiver. In *7th Management Comitee/Working Group Meeting and Workshop of COST ic0803*, Sep. 2011.

[10] L. Zwirello, C. Heine, X. Li, and T. Zwick. An UWB Correlation Receiver for Performance Assessment of Synchronization Algorithms. In *IEEE International Conference on Ultra-Wideband (ICUWB)*, pages 210–214, Sep. 2011.

[11] L. Zwirello, X. Li, E. Pancera, and T. Zwick. An Ultra-Wideband Communication System with Localization Capabilities - Hardware Realization of HF-Frontend Modules. In *6th Management Comitee/Working Group Meeting and Workshop of COST ic0803*, Apr. 2011.

[12] L. Zwirello, P. Pahl, Ch. Heine, and T. Zwick. Status Report on the FCC-Compliant UWB Transceiver Implementation. In *6th IEEE UWB Forum on Sensing and Communication*, May 2011.

[13] L. Zwirello, M. Harter, H. Berchtold, J. Schlichenmaier, and T. Zwick. Analysis of the Measurement Results Performed With an Ultra-Wideband Indoor Locating System. In *7th German Microwave Conference (GeMiC)*, Mar. 2012.

[14] L. Zwirello, C. Heine, X. Li, T. Schipper, and T. Zwick. SNR Performance Verification of Different UWB Receiver Architectures. In *European Microwave Conference (EuMC)*, pages 1316–1319, Nov. 2012.

[15] L. Zwirello, M. Hesz, Y.L. Sit, and T. Zwick. Algorithms for Synchronization of Coherent UWB Receivers and their Application. In *IEEE International Conference on Ultra-Wideband (ICUWB)*, pages 207–211, Sep. 2012.

[16] L. Zwirello, M. Janson, X. Li, S. Dosch, M. Schneider, J. Fleischer, and T. Zwick. Ultra-Breitbandsysteme für die flexible, adaptive Steuerung von Produktionsanlagen. Projektbericht für die Baden-Württemberg Stiftung gGmbH, Ministerium für Wissenschaft, Forschung und Kunst, Jun. 2012.

[17] L. Zwirello, L. Reichardt, X. Li, and T. Zwick. Impact of the Antenna Impulse Response on Accuracy of Impulse-Based Localization Systems. In *6th European Conference on Antennas and Propagation (EuCAP)*, pages 3520–3523, Mar. 2012.

[18] L. Zwirello, T. Schipper, M. Harter, and T. Zwick. UWB Localization System for Indoor Applications: Concept, Realization and Analysis. *Journal of Electrical and Computer Engineering*, 2012:1–11, May 2012.

[19] L. Zwirello, T. Schipper, M. Jalilvand, and T. Zwick. An IR-UWB Indoor Localization System with Sub-Decimeter Accuracy. In *8th Management Comitee/Working Group Meeting and Workshop of COST ic0803*, May 2012.

[20] L. Zwirello and T. Zwick. Indoor Localization via Impulse-Radio Ultra-Wideband: from Theory to Application. In *IEEE/MTT-S International Microwave Symposium (IMS), Workshop on Wireless Positioning and Tracking in Indoor/Urban Environments*, Jun. 2012. Invited Talk.

[21] L. Zwirello, C. Ascher, X. Li, S. Werling, G.F. Trommer, and T. Zwick. Sensor Data Fusion in UWB-Supported Inertial Navigation Systems for Indoor Navigation. In *IEEE International Conference on Robotics and Automation (ICRA)*, pages 1–6, May 2013.

[22] L. Zwirello. *Ultra-Wideband RF System Engineering*, chapter UWB applications, pages 138–161. Cambridge University Press, 2013.

[23] L. Zwirello, T. Schipper, S. Dosch, J. Fleischer, and T. Zwick. *E&E Kompendium 2013*, chapter Mobile Arbeitsmaschinen simultan orten und ansteuern, pages 267–270. publish-industry Verlag GmbH, 2013.

Co-authored publications

[24] E. Pancera, L. Zwirello, T. Zwick, and W. Wiesbeck. Spectrum Optimization in Ultra Wideband Systems. In *3rd Management Comitee/Working Group Meeting and Workshop of COST ic0803*, Oct. 2009.

[25] J. Timmermann, L. Zwirello, G. Adamiuk, and T. Zwick. Performance of Conventional and Optimal Pulse Shapes in Non-ideal UWB Transmission. In *COST 2100, TD(09)707*, Feb. 2009.

[26] J. Timmermann, L. Zwirello, E. Pancera, Malgorzata Janson, and T. Zwick. Comparing Non-Ideal Ultra-Wideband Transmission for European and FCC Regulations. In *European Radar Conference (EuRAD)*, pages 351–354, Sep. 2009.

[27] G. Adamiuk, L. Zwirello, S. Beer, L. Reichardt, and T. Zwick. Extension of Monopulse-Radar-Technique to UWB-Systems. In *COST 2100, TD(10)12089*, Nov. 2010.

[28] G. Adamiuk, L. Zwirello, S. Beer, and T. Zwick. Omnidirectional dual-orthogonal, linearly polarized UWB Antenna. In *European Microwave Confeence (EuMC)*, pages 854–857, Sep. 2010.

[29] G. Adamiuk, L. Zwirello, L. Reichardt, and T. Zwick. UWB Cross-Polarization Discrimination with Differentially Fed, Mirrorred Antenna Elements. In *IEEE International Conference on Ultra-Wideband (ICU-WB)*, pages 1–4, Sep. 2010.

[30] X. Li, E. Pancera, L. Zwirello, H. Wu, and T. Zwick. Detection of Water Accumulation in the Human Bladder with Ultra Wideband Radar. In *COST 2100, TD(10)10012*, Feb. 2010.

[31] X. Li, E. Pancera, L. Zwirello, H. Wu, and T. Zwick. Ultra Wideband Radar for Water Detection in the Human Body. In *German Microwave Conference (GeMiC)*, pages 150–153, Mar. 2010.

[32] E. Pancera, X. Li, L. Zwirello, and T. Zwick. Performance of Ultra Wideband Antennas for Monitoring Water Accumulation in Human Bodies. In *European Conference on Antennas and Propagation (EuCAP)*, pages 1–5, Apr. 2010.

[33] E. Pancera, X. Li, L. Zwirello, T. Zwick, and W. Wiesbeck. UWB Radar Systems for Medical Dagnostics: Applications and Measurements Results. In *4th Management Comitee/Working Group Meeting and Workshop of COST ic0803*, Feb. 2010.

[34] E. Pancera, L. Zwirello, T. Zwick, and W. Wiesbeck. Quantification of the Impact of the Antenna Non-Idealities in UWB Transmission Systems. In *IEEE Antennas and Propagation Society International Symposium (APS)*, pages 1–4, Jul. 2010.

[35] C. Ascher, L. Zwirello, T. Zwick, and G.F. Trommer. Integrity Monitoring for UWB/INS Tightly Coupled Pedestrian Indoor Scenarios. In *International Conference on Indoor Positioning and Indoor Navigation (IPIN)*, pages 1–6, Sep. 2011.

[36] X. Li, M. Jalilvand, L. Zwirello, and T. Zwick. Array Configurations of a UWB Near Field Imaging System for the Detection of Water Accumulation in Human Body. In *European Radar Conference (EuRAD)*, pages 170–173, Oct. 2011.

[37] X. Li, M. Jalilvand, L. Zwirello, and T. Zwick. Synthetic Aperture-Based UWB Imaging System for Detection of Urine Accumulation in Human Bladder. In *IEEE International conference on Ultra Wideband (ICUWB)*, pages 351–354, Sep. 2011.

[38] J. Pontes, G. Adamiuk, S. Beer, L. Zwirello, X. Li, and T. Zwick. Novel Design Method for Frequency Agile Beam Scanning Antenna Arrays. In *IEEE International Workshop on Antenna Technology (iWAT)*, pages 424–427, Mar. 2011.

[39] H. Wu, L. Zwirello, X. Li, L. Reichardt, and T. Zwick. Motion Compensation with One-axis Gyroscope and Two-axis Accelerometer for Automotive SAR. In *German Microwave Conference (GeMiC)*, pages 1–4, Mar. 2011.

[40] C. Ascher, L. Zwirello, C. Hansmann, S. Werling, T. Zwick, and G.F. Trommer. Radio-Asissted Inertial Navigation System by Tightly Coupled Sensor Data Fusion: Experimental Results. In *International Conference on Indoor Positioning and Indoor Navigation (IPIN)*, pages 1–7, Nov. 2012.

[41] M. Harter, T. Schipper, L. Zwirello, A. Ziroff, and T. Zwick. 24 GHz Digital beamforming radar with T-shaped antenna array for three-dimensional object detection. *International Journal of Microwave and Wireless Technologies*, 4:327–334, Jun. 2012.

[42] M. Harter, L. Zwirello, T. Schipper, A. Ziroff, and T. Zwick. A Novel Sensing Method for Automatic Guidance of Trolley Vehicles Based on Digital Beamforming Radar. In *IEEE International Conference on Wireless Information Technology and Systems (ICWITS)*, pages 1–4, Nov. 2012.

[43] X. Li, M. Jalilvand, J. Yan, L. Zwirello, and T. Zwick. A Multiband Slotted Bowtie Antenna for Stroke Detection. In *IEEE International conference on Ultra Wideband (ICUWB)*, pages 439–442, Sep. 2012.

[44] X. Li, Y. Sit, L. Zwirello, M. Jalilvand, and T. Zwick. An E-Field Probe Based Near-Field Measurement System for On- and In-Body Antennas. In *European Microwave Conference (EuMC)*, pages 297–300, Oct. 2012.

[45] X. Li, L. Zwirello, M. Jalilvand, and T. Zwick. Design and Near-Field Characterization of a Planar On-body UWB Slot-Antenna for Stroke Detection. In *IEEE International Workshop on Antenna Technology (iWAT)*, pages 201–204, Mar. 2012.

[46] L. Reichardt, J. Kowalewski, L. Zwirello, and T. Zwick. Compact, Teflon Embedded, Dual-Polarized Ultra Wideband Antenna. In *Antennas and Propagation Society International Symposium (APS/URSI)*, pages 1–2, Jul. 2012.

[47] T. Schipper, M. Harter, L. Zwirello, T. Mahler, and T. Zwick. Systematic Approach to Investigate and Counteract Interference-Effects in Automotive Radars. In *European Radar Conference (EuRAD)*, pages 190–193, Nov. 2012.

[48] M. Harter, T. Schipper, L. Zwirello, A. Ziroff, and T. Zwick. Detection of Overhead Contact Lines with a 2D-Digital-Beamforming Radar System for Automatic Guidance of Trolley Trucks. *International Journal of Vehicular Technology*, 2013:1–5, Jan. 2013.

[49] M. Jalilvand, X. Li, L. Zwirello, and T. Zwick. Sensititvity Analysis of UWB Minimum Variance Beamformer for Medical Imaging. In *European Radar Conference (EuRAD)*, Oct. 2013.

[50] X. Li, Y.L. Sit, L. Zwirello, and T. Zwick. A Miniaturized UWB Stepped-Slot Antenna for Medical Diagnostic Imaging. *Microwave and Optical Technology Letters*, 55(1):105–109, Jan. 2013.

[51] T. Schipper, S. Prophet, L. Zwirello, M. Harter, L. Reichardt, and T. Zwick. Simulation Framework for the Estimation of Future Interference Situations between Automotive Radars. In *Antennas and Propagation Society International Symposium (APS/URSI)*, Jul. 2013.